Clinical Nutrition Case Studies

Third Edition

Wayne Billon
University of Southern Mississippi

West/Wadsworth
I(T)P® An International Thomson Publishing Company

Belmont, CA • Albany, NY • Boston • Cincinnati • Johannesburg • London • Madrid• Melbourne
Mexico City • New York • Pacific Grove, CA • Scottsdale, AZ • Singapore • Tokyo • Toronto

Nutrition Editor: Peter Marshall
Development Editor: Laura Graham
Assistant Editor: Tangelique Williams
Editorial Assistant: Keynia Johnson
Marketing Manager: Becky Tollerson
Project Editor: Howie Severson
Print Buyer: Barbara Britton
Cover Design: Carole Lawson
Printer: Mazer

Printed in the United States of America
1 2 3 4 5 6 7 8 9 10

For more information, contact Wadsworth Publishing Company, 10 Davis Drive, Belmont, CA 94002, or electronically at http://www.wadsworth.com

International Thomson Publishing Europe
Berkshire House
168-173 High Holborn
London, WC1V 7AA, United Kingdom

Nelson ITP, Australia
102 Dodds Street
South Melbourne
Victoria 3205 Australia

Nelson Canada
1120 Birchmount Road
Scarborough, Ontario
Canada M1K 5G4

International Thomson Publishing Southern Africa
Building 18, Constantia Square
138 Sixteenth Road, P.O. Box 2459
Halfway House, 1685 South Africa

International Thomson Editores
Seneca, 53
Colonia Polanco
11560 México D.F. México

International Thomson Publishing Asia
60 Albert Street
#15-01 Albert Complex
Singapore 189969

International Thomson Publishing Japan
Hirakawa-cho Kyowa Building, 3F
2-2-1 Hirakawa-cho, Chiyoda-ku
Tokyo 102, Japan

ISBN 0-534-54657-9

 This book is printed on acid-free recycled paper.

TABLE OF CONTENTS:

PREFACE

One day, about nine years ago, I was sitting in my office making out a case study to use as a test for my clinical nutrition class when a young salesman from West Publishing walked in and asked if there was anything by way of nutritional textbooks I needed. I replied, "A case study book." I explained that I had used case studies in my classes for years and would like to see an updated book of case studies. His reply was simple: "Why don't you write one?" At first I chuckled, but as the days passed and I considered the idea, I decided it was a possibility. Now the third edition has been printed. The intent has not changed: to prepare students in clinical dietetics for clinical practice.

Like the second edition, the third edition has been greatly revised. Abbreviations, medications, laboratory values, and formulas have been added or updated. In particular, the diabetes cases have been updated to reflect the new standards, guidelines, and terminology. New cases on this subject have been added. There is also a new appendix of available WEB sites. These changes were made to keep abreast of the times. Almost all other changes are a based on comments and suggestions made by reviewers. For example, the student will no longer have to look up the normal value for the laboratory tests. The norms are listed in the tables when the labs are presented. This will save the student time and is consistent with the "real" clinical chart. SOAP notes and hand written doctors orders and progress notes are additional new features.

Another request was for references. In this edition, each case study is followed by a list of related references. Some of the references were used in the case study and others are included for additional information. Reviewers recommended the inclusion of additional cases in the areas of pediatrics, gestational diabetes, and peritoneal dialysis. A total of five new cases have been added. Some of the cases have been revised to such an extent that they are like new cases. The use of herbs and nutritional supplements has been incorporated into some of the cases also. This manual has been recognized so that most of the easier cases can be found at the beginning of the text.

Another major change involves the use of ideal body weight versus reference body weight. The Hamwi formula is a quick method commonly used by dietitians to determine someone's "ideal body weight." The term "ideal" has been questioned because, with changing weight standards and the possible significant variation between individuals, the "ideal weight" is difficult to define. *The Dietary Guidelines for Americans*, revised in 1995, suggests a new weight range for all ages and both genders. The lower end of the range is intended for less muscular individuals and the higher end for those who are more muscular. The new dietary guidelines are used throughout this book with the "midpoint" of the range being referred to as the "reference body weight" (RBW). The RBW can be adjusted from the midpoint to anywhere in the range depending on the muscularity of the individual being assessed. The weight ranges recommended by *The Dietary Guidelines for Americans*, along with the midpoints, can be found in **Table A - 9 of Appendix A**. If you prefer, you may continue to use the IBW as described by Hamwi in **Table - 8 of Appendix A**. In each case study you can use one or the other or both and compare the differences. This change is based on suggestions made by reviewers.

One of the most difficult elements of teaching clinical nutrition is presenting the material in such a way that students can relate the classroom instruction to the clinical setting. In their first clinical course, students need to be introduced to medical terminology, abbreviations, therapeutic diets, laboratory values, assessment techniques, interviewing techniques, chart reading and recording, disease states as they relate to diet, and more. The information collected has to be coordinated into an appropriate recommendation or observation that can be concisely written into an organized chart note. This is true for diet technicians, dietitians, nurses, respiratory therapists, and medical students. It is an overwhelming task for both the student and instructor. However, if realistic cases can be presented in the classroom setting, the students are more apt to relate the principles to practice when they enter the clinical setting. Having taught clinical dietetics for seventeen years in the college classroom and six years in a hospital setting, I have observed that lack of confidence accompanied by an overwhelming amount of information is a major deterrent to students being able to relax and function at their fullest potential. I believe this lack of confidence can be effectively reduced with the case studies in this book. The cases are realistic and are based on true-life situations. The completion of the case studies before going into the clinical facility will give the student a broader knowledge base and a greater level of confidence.

Some case studies are long and may include questions not covered in class. Those questions may either be omitted, researched by the students as additional projects, or discussed by the instructor. Some questions may have to be omitted because of the level of difficulty, and questions may need to be added to cover material presented in class. **The studies were designed to cover the main points and are not intended to cover everything.** This case study book is not intended to be a main text but is a supplement to a course textbook. Most cases cover more than one topic in an attempt to include a greater variety of situations. I hope that the studies will stimulate thought that *will generate additional questions.* It is at the discretion of the instructor to add to or delete from the case studies to meet the needs of the students using this book.

If, by writing this book, I can impart some knowledge or understanding that will help someone become a better person and be able to more efficiently help someone else, I will be fulfilled. To the person studying this book, I hope you will enjoy learning about nutrition as much as I enjoy teaching nutrition.

Wayne E. Billon
December 14, 1998

ACKNOWLEDGMENTS

Producing this book was no easy task. The third edition took much more effort than either of the previous editions. I received assistance and encouragement from many individuals, including my students, who added comments as they completed case study assignments. These comments, along with those of several reviewers, gave insight into the development of the third edition. To all of these I give my most sincere thanks.

I particularly want to thank my loving wife, Cathy, for her encouragement and understanding, as well as her typing skills. This edition was as hard on her as it was on me. Peter Marshall and Laura Graham of West/Wadsworth Publishing have been extremely helpful not only in the coordination of the editing, reviewing, and publishing, but also with their insight and encouragement. Many thanks to my director, Dr. Anita Stamper, for her assistance and guidance. Many of the improvements in the third edition are the result of ideas I received from reviewers. For this I am most grateful and would like to extend a special thank you to the following reviewers:

Sara Long Anderson, Southern Illinois University at Carbondale
Susan Appelbaum, St. Louis Community College - Florissant Valley
Erin Bettinger, SUNY College of Technology and Agriculture
Dorothy Chen-Maynard, California State University - San Bernardino
Barbara Cosper, Western Carolina University
Judy Gaare, New York Institute of Technology
Janet W. Gloeckner, James Madison University
Jeffrey Harris, West Chester University
Betsy Holli, Rosary College
Mary Hubbard, Grossmont College
Mardell Wilson, Illinois State University

INDEX TO OTHER NUTRITION TEXTBOOKS

Billon	Whitney Caltaldo Rolfes	Cataldo DeBruyne Whitney	Zeman	Williams	Williams	Davis Sherer	Weigley Mueller Robinson	Mahan Escott-Stump
Clinical Nutrition Case Studies 3E	Under-standing Normal and Clinical Nutrition 5E	Nutrition and Diet Therapy 5E	Clinical Nutrition and Dietetics 2E	Essentials of Nutrition and Diet Therapy 8E	Nutrition and Diet Therapy 7E	Applied Nutrition and Diet Therapy for Nurses 2E	Robinson's Basic Nutrition and Diet Therapy	Krause's Food, Nutrition, & Diet Therapy 9E
Case Study	Chapter	Chapter	Chapter	Chapter	Chapter	Chapter	Chapter	Chapter
1	6	4	18	5	5	3	7?	16
2	8	10	- - -	17	14	10	8	22
3,4	1,8,9,15, 16,17	15,16,17	3,18	18	15	11	3,16	17,18,19
5	9	9	3,12,18	7	6	20	9	13
6	20	14	4,10,19	16,19	13,16	16,20,23, 24	15,18	18,14
7,9,10	8,9	9	12	7,12	6,9	17,19	2,9	21
8	21	18	12	23,28	19,22	19	21,22	21
11,12,15, 16	21	18	7	23,28	18,22	20,25	21,22	27
13,14	- - -	- - -	- - -	- - -	- - -	- - -	- - -	11
17,18	22	19	8	23	18,22	19,25	22	28
19,32,33	26	28	5,8,13	24	18	13,26	23	29
20	26	20,28	8	24	18	26	23	29?
21,22,23, 24	28	26	5,10	25	19,22	24	24	23,33
25,26,27, 28,29	27	25	5,9,11	26	20	27	26	31
30,31	25	23,24	5,19	31	-----	13	- - -	30
34,35	24,30	22,24	5,15,17	21,30	18,25	13,21,22, 25	17,25	36
36	30	24	6	29	23	22	29	37
37,38,39	28,29	26,27	9,10	25,27	19,21	24,29	24,28	24,35
40,41	22,24	19,22	5,8	21,23	18,17	13,19,25	17,22	28
42	24,25,	22,23	5,14	21	22	13,21	17,21	30

CASE STUDY #1
VEGETARIANISM

INTRODUCTION

The purpose of each of the case studies is to gradually introduce the student to one or more of the following processes: reading and understanding a patient's chart, interviewing a patient, interpreting pertinent data, and assimilating that data into an appropriate nutritional care plan. The first study is a warm-up that includes a small number of abbreviations and lab values. The cases will become more difficult as the book progresses. This study is about a person who is a true vegan and is concerned about environmental issues. Review information about vegetarian diets, possible deficiencies, and complications. The reference lists at the ends of the chapters are a good place to start to look for information about the cases. Most answers should appear in a basic or an advanced clinical nutrition textbook.

SKILLS NEEDED

ABBREVIATIONS:

Knowledge of the following abbreviations is required in order to understand this case. You should learn these abbreviations before you begin to read the study.

Hct : hematocrit
Hgb : hemoglobin
MCV : mean corpuscular volume

RBW : reference body weight
Ser Alb : serum albumin
YOWF : year old white female

LABORATORY VALUES:

An ability to interpret the nutritional significance of the following laboratory values will be needed for this case study: Hct, Hgb, MCV, and Ser Alb (Appendix B).

FORMULAS:

The formulas used in this case study are for reference body weight and percent reference body weight. The formulas can be found in Appendix A, Tables A - 7 through 10.

SR is a 23 YOWF college student who is very much concerned about the pollution of the environment. She believes mankind should do everything possible to preserve nature, including protecting the wildlife. SR is very health conscious and advocates that everyone should be a vegetarian. She is a true vegan and has been for two and one-half years. She is 5'2" and weighs 98 lbs. She eats absolutely no animal products. Her diet includes tofu, alfalfa sprouts, and legumes of various types such as soybeans, lentils, black beans, chick peas, and peanuts. She drinks canned soy milk. The only vegetable oils that she uses are safflower oil or canola oil. She does not take a vitamin or mineral supplement on a regular basis because she believes they are not necessary. She believes she can obtain all of her vitamins and minerals from the foods she eats. However, when she is extremely stressed out, she takes a stress vitamin and mineral supplement. The supplement has to be an "all-natural organic" vitamin and mineral mix. SR has not felt stressed in a long time so she has not been taking her vitamin and mineral tablets. She takes no other nutritional supplement, protein drink, or mineral supplement, such as calcium,

Tums, etc. She does not smoke or consume alcoholic beverages. SR eats only "natural" foods. She does not eat bleached flours but uses whole grain cereals and bread. SR makes most of the bread she eats and, as with everything she prepares herself, uses only "all-natural" ingredients. She refuses to use a microwave or to consume previously prepared foods. She takes in a variety of fruits and vegetables. SR eats rice or some grain with every meal that has legumes.

SR has been feeling progressively weaker over the last year. She is not very active and gets little exercise. In spite of her health consciousness, she is a very sedentary person. Her appearance is very pale, her hair lacks luster, she has pale gums, and her nail beds take several seconds to return to their normal pink color when depressed. If she uses the stairs to get to her office, the climb up three flights causes her to exhibit shortness of breath.
****

QUESTIONS:

1.　　Define the following terms:

　　　Vegan:

　　　Ovo-vegetarian:

　　　Lacto-vegetarian:

　　　Pesco-vegetarian:

　　　Pollo-vegetarian:

　　　Natural vitamins:

　　　Organic vitamins:

　　　Synthetic vitamins:

Information Box 1-1
The Hamwi formula[a] has commonly been used by dietitians to determine someone's "ideal" body weight quickly. The term "ideal" is being questioned by many practitioners, educators, and researchers because, with weight standards changing and the possible significant variation between individuals, the meaning of "ideal" is difficult to define. *The Dietary Guidelines for Americans*[b], revised in 1995, suggested a weight range for all ages and both genders. The lower end of the range is intended for less muscular individuals and the higher end of the range for those who are more muscular. The new dietary guidelines will be used throughout this book with the midpoint of the range being referred to as the "reference body weight" (RBW). The new guidelines for weight ranges, along with the midpoint of the ranges, can be found in Appendix A, Table A - 9. If this is not acceptable and the Hamwi formula is desired, directions on its use can be found in Appendix A, Table A - 8.
[a] Miller, M.A. (1985). A calculated method for determination of ideal body weight. Nutritional Support Services pp. 31-33.
[b] *Report of the Dietary Guidelines Advisory Committee on the Dietary Guidelines for Americans.* (1995). Washington, D.C.: Government Printing Office.

2.   Determine SR's RBW and the percent of her RBW (Appendix A - 7 through 10).

3.   What kind of oils are safflower and canola oils? What is another name for canola oil?

4.   SR likes tofu, alfalfa sprouts, legumes, lentils, chick peas, and soy milk. What are these foods good sources of, if anything?

5.   What are the sources of calcium in SR's diet?

6.   What are the sources of complete protein in SR's diet?

7.   What possible nutritional deficiencies could SR have based on the information given?

8.   Do you agree or disagree with SR's theory about vitamin/mineral supplements for vegans? Explain.

9.   Does stress increase our need for vitamins and minerals? Explain.

10.   Which type of vitamin functions the best in the human body—natural, organic, or synthetic? Explain.

11.   Are "all-natural" or organically grown foods healthier than processed foods?

12.   Is irradiation safe? Explain.

13. Predict what the lab values would be for Hct, Hgb, MCV, Ser Alb, and percent lymphocytes using the table below. Based on the information given, do you believe they would be below normal, above normal, or in the normal range? Give reasons why you would make these predictions.

Test	Predict: High Normal Low	Normal Range	Reason for Prediction:
Hct			
Hgb			
MCV			
Ser Alb			
Lymphocytes			

14.   If you were counseling SR about her diet and she insisted on remaining a vegan, what recommendations would you give her to improve her nutritional intake? What animal products might she include in her diet that the consumption of which would actually help preserve the animals from which the products are derived?

15.   Besides those that are in SR's diet, name three sources of complementary proteins.

16.   The clinical evaluation of SR included paleness, dull hair coat, nail beds that did not return to normal color rapidly after being depressed, and pale gums. What do these clinical signs suggest?

Related Reading

1.   Fischbach, F.T. (1995). A Manual of Laboratory & Diagnostic Tests. 5th Ed. Philadelphia. J.B. Lippincott Company.

2.   Food and Nutrition Board, National Research Council. (1989) *Recommend Dietary Allowances,* 10th Ed. Washington: National Academy Press, pp. 52-77.

3.   Position paper of the American Dietetic Association. (1995): Biotechnology and the future of food. *J. Am. Diet. Assoc.* 95:1429.

4.   Position paper of the American Dietetic Association. (1996): Food irradiation. *J. Am. Diet. Assoc.* 96:69.

5.   Position paper of the American Dietetic Association. (1997): Vegetarian diets. *J. Am. Diet. Assoc.* 97:1317.

6.   Position paper of the American Dietetic Association. (1997): Natural resource conservation and waste management. *J. Am. Diet. Assoc.* 97:425.

7.   Position paper of the American Dietetic Association. (1994): Enrichment and fortification of foods and dietary supplements. *J. Am. Diet. Assoc.* 94:661.

8.   Position paper of the American Dietetic Association. (1996): Vitamin and mineral supplementation. *J. Am. Diet. Assoc.* 96:73.

9.   Prendergast, A. & Fulton, F.L. (1997). Medical Terminology: A Text/Workbook. 4th Ed. Redwood City, California. Addison-Wesley Nursing.

10. Statement of FANSA[1] (1997): What does the public need to know about dietary supplements? *J. Am. Diet. Assoc.* 97:728.

11. Walter, P. (1997). Effects of vegetarian diets on aging and longevity. *Nutrition Reviews.* International Life Sciences Institute, Washington, DC. 55:1, pp. S61-S68.

12. Whitney, E.N., Cataldo, C.B. & Rolfes, S.R. (1998). *Understanding Normal and Clinical Nutrition,* 5th Ed. West/Wadsworth.

13. Zeman, F.J. & Ney, D.M. (1996). Applications in Medical Nutrition Therapy, 2nd Ed. pp. 116-123. Merrill/Prentice Hall. 1996.

[1] The Food and Nutrition Science Alliance, a partnership of four professional scientific societies: American Dietetic Association, American Society for Clinical Nutrition, American Society for Nutritional Sciences, and the Institute of Food Technologists.

CASE STUDY #2
NUTRITION AND BODY-BUILDING

INTRODUCTION

This study concerns a young man who is interested in building muscle mass and believes everything he hears about protein supplements. Assume that this young man, whom we will call RJ, is working out at the same health club that you belong to. You meet RJ at the club and he finds out that you are a registered dietitian. He has several questions to ask you about nutrition.

SKILLS NEEDED

ABBREVIATIONS:

Knowledge of the following abbreviations is required in order to understand this case. You should learn these abbreviations before you begin to read the study.

AA : amino acids
BIA : bioelectrical impedance
BMI : body mass index
g : gram
lbs : pounds

RBW : reference body weight
RD : Registered Dietitian
TSF : triceps skinfolds
YOWM : year old white male
µg : microgram

FORMULAS:

The formulas used in this case study are reference body weight, percent reference body weight (Appendix A, Tables A - 7 through A - 10), body mass index (Appendix A, Tables A - 12 and 13), and the Harris-Benedict equation (Appendix D, Table D - 1).

RJ is a 21 YOWM who has a great interest in body building. He is a full-time college student and spends all of his free time at a health club working out with free-weights. His ambition is to compete in body building at the local and state levels and to one day be state champion. In the back of his mind, he is also thinking of national competition. He is 6'1" and weighs 230 lbs., but because of his lean body mass, does not look overweight. He works out every day for two to three hours, working different body parts each time. His goal is to gain 25 more pounds. The health club he belongs to has state-of-the-art equipment and supportive services. Among the supportive services are cardiovascular and respiratory evaluation, fat analysis, and dietary instruction. One day you have the opportunity to meet RJ and talk with him at the health food bar in the health club. He finds out that you are a registered dietitian and asks you questions about nutrition. He relates the above information to you and tells you that he has taken some extra supplements and wants to know what you think. He also tells you that he already talked to the RD who works at the health club but because her main job is working with people who are overweight, he does not have confidence in her.

RJ tells you that he is on a high-protein, moderate-carbohydrate, low-fat diet. He has obtained a calorie-counter book from a newsstand and uses it to determine the amounts of carbohydrate, fat and protein he ingests. He explained that he takes in 10 percent or less of his total calories from fat. He is taking 200 g of protein per day. He understands that the best form of protein is eggs, so his 200 g of protein includes four raw eggs per day. In addition, he

takes 3 g of a powdered amino acid supplement before working out, 3 g after working out, and 3 g in the morning and evening. The AA supplement is particularly high in lysine, arginine, and ornithine and is added to milk. He is also taking a high-stress multiple vitamin and mineral tablet every day. The tablet contains 200 percent of his requirement for vitamins A and D. The vitamin A is in the acetate form. What he knows about nutrition comes from reading various body building magazines. In some of his magazines he has read that vitamin B_6 is necessary for protein metabolism, so he takes 1 g of vitamin B_6 every day to help synthesize protein at a faster rate. He has also read that vitamin C is necessary for protein synthesis, so he takes 2 g of vitamin C per day. Both are in addition to the vitamin B_6 and vitamin C that are in the multiple vitamin he is taking.

The health club also has the means to determine body fat percentage by TSF, bioelectrical impedance (BIA), or underwater weighing. RJ was very much interested in knowing his percent body fat, so he had the BIA test done. The results indicated that RJ had 10 percent body fat. RJ was disappointed with this and wanted to get his body fat down to at least 7 percent.
****

QUESTIONS:

1.   Determine RJ's RBW and percent of RBW (Appendix A, Tables A - 7 through 10).

2.   Would you classify RJ as within his normal range of weight, overweight, or obese?

3.   If you classified RJ as overweight, would you use an adjusted body weight formula or his actual weight to calculate his energy needs (Appendix A, Table A - 11)? Explain your answer.

4.   Calculate RJ's BMI (Appendix A - 12).

5.   Based on his BMI, how would you classify RJ (Appendix A - 13)? Do you believe this classification is accurate? Briefly explain your answer.

6.   Using the Harris-Benedict equation (Appendix D, Table D -1), calculate RJ's energy needs. Refer to the appropriate activity factor to determine his total energy requirement for a day (see Appendix D - 2).

7.   Are RJ's protein needs different because he is a weight lifter and has more muscle than the average male his age? In calculating his protein needs, would you use his RBW, actual body weight, or adjusted body weight to make your determinations? Explain your answer.

8.   How many grams of protein per kilogram of body weight is RJ taking in? What will happen to the extra protein?

9.   What is the advantage or disadvantage of RJ taking in the extra protein supplements?

10.   Considering RJ's high protein diet, the four raw eggs per day, and the protein powder, how would you advise RJ concerning his protein intake?

11.   What would tell RJ about eating four raw eggs per day?

12.   Concerning RJ's carbohydrate intake, discuss the following:

  • Compare his actual carbohydrate intake with his carbohydrate requirement.

  • Compare the amount of energy obtained from carbohydrate versus from fat in the average person to that of an athlete/weightlifter.

  • Where is RJ getting most of his energy in his current diet? What are the advantages and/or disadvantages of this energy supply?

13.   Does RJ need extra vitamin B$_6$ and/or vitamin C? If he does, how much does he need?

14.   Will the extra doses of vitamins B_6 and C produce the results RJ is expecting? Explain your answer.

15.   What harmful effects could result from excessive intake of vitamins B_6 or C?

16.   What is your recommendation to RJ concerning the multiple vitamin that he is taking?

17.   What advice would you give RJ about his percent body fat and his goal of trying to reach seven percent or less body fat?

18. Describe and compare the following three methods of determining percent body fat in relation to techniques, accuracy of each test, time involved, and disadvantages.

TSF	BIA	UNDERWATER WEIGHING

As time goes by, RJ comes to you again for additional advice about the latest fad, creatine. In a conversation in the gym, RJ heard about several new supplements his friends called "ergogenic aids." It seems everyone at the gym is taking creatine monohydrate to help them "bulk-up." RJ says it really has helped the guys that took it so he bought some and is ready to start a loading phase. The directions indicate a loading dose of 15 to 20 g for 5 days and then a maintenance dose of 5 g per day. RJ has decided that if this dose helped his friends, twice that amount will help him twice as much. He wants to know what you think about creatine monohydrate and the recommended doses. RJ has also heard about chromium picolinate and is planning to take 200 µg per day. He is very interested in your opinion of this supplementation.

19.   What is the body's source of creatine? Describe its function in the production of energy in muscle.

20.   What does the current literature say about the value of dietary supplementation of creatine monohydrate?

21.   How would you advise RJ about the supplementation of creatine, and particularly concerning the doubling of the recommended dose?

22.   Define "ergogenic aid" and give several specific examples.

23.   What is chromium picolinate?

24. What is the human requirement for chromium and what does it do in the body?

25. What is the basis of the theory for dietary supplementation of chromium picolinate?

26. What does the current literature say about the value of dietary supplementation of chromium?

27. How would you advise RJ about the supplementation of chromium?

Related Reading

1. Clarkson, P.M. Nutritional Supplements for weight gain. (1998). *Gatorade Sports Science Institute.* 11(1).

2. Clancy, S.P., Clarkson, P.M., DeCheke, M.E., Nosaka, K., Freedom, P.S., Cunningham, J.J., & Valentine, B. (1994). Effects of chromium picolinate supplementation on body composition, strength, and urinary chromium loss in football players. *Int. J. Sport Nutr.* 4:142-153.

3. Eichner, E.R. (1997). Ergogenic Aids: What athletes are using and why. *Phys. Sportsmed.* 25(4):70-83.

4.   Evans, G.W. (1989). The effects of chromium picolinate on insulin controlled parameters in humans. *Int. J. Biosoc. Med. Res.* 11:163-180.

5.   Flisinska-Bojanowska, A. (1996). Effects of oral creatine administration on skeletal muscle protein and creatine levels. *Biol. Sport.* 13:39-46.

6.   Fogelholm, G.M., Näveri, H.K., Kiilavuori, K.T.K., & Härkönen, M.H.A. (1993). Low-dose amino acid supplementation: no effect on serum human growth hormone and insulin in male weightlifters. *Int. J. Sports Nutr.* 3:290-297.

7.   Food and Nutrition Board, National Research Council. (1989). *Recommend Dietary Allowances,* 10th Ed. Washington: National Academy Press, pp. 52-77.

8.   Hallmark, M.A., Reynolds, T.H., DeSouza, C.A., Dotson, C.O., Anderson, R.A., & Rogers, M.A. (1996). Effects of chromium and resistive training on muscle strength and body composition. *Med. Sci. Sports Exerc.* 28:139-144.

9.   Hasten, D.L., Rome, E.P., Franks, B.D., & Hegsted, M. (1992). Effects of chromium picolinate on beginning weight training students. *Int. J. Sport Nutr.* 2:343-350.

10.   SCAN's Guide to Nutrition and Fitness Resources: SCAN'S PULSE, supp., (1997). *The American Dietetic Association.* Chicago, Il.

11.   Jacobs, I., Bleue, S., & Goodman, J. (1997). Creatine ingestion increases anaerobic capacity and maximum accumulated oxygen deficit. *Can. J. Appl. Physiol.* 22:231-243.

12.   Lamb, R.D. & Wardlaw, G.M. (1991). Sports Nutrition. *Nutri-News.* Mosby-Year Book.

13.   Lambert, M.J., Hefer, J.A., Millar, R.P., & Macfarlane, P.W. (1993). Failure of commercial oral amino acid supplements to increase serum growth hormone concentrations in male body-builders. *Int. J. Sports Med.* 3:298-305.

14.   Lee, R.D. & Nieman, D.C. (1993). Nutritional Assessment. Brown and Benchmark, pp. 137-153.

15.   Lefavi, R.G., Anderson, R.A., Keith, R.E., Wilson, G.D., McMillan, J.L., & Stone, M.H. (1992) Efficacy of chromium supplementation in athletes: emphasis of anabolism. *Int. J. Sport Nutr.* 2:111-122.

16.   Lemon, P.W.R. (1995). Do athletes need more dietary protein and amino acids? *Int. J. Sport Nutr.* 5:S39-S61.

17.   Lemon, P.W.R. (1994). Are dietary protein needs affected by regular exercise? *Insider* 2(3):October.

18.   Lukaski, H.C., Bolonchuk, W.W., Siders, W.A., & Milne, D.B. (1996). Chromium supplementation and resistance training: effects of body composition, strength, and trace element status of men. *Am. J. Clin. Nutr.* 63:954-965.

19.   Mertz, W. (1992). Chromium: History and nutritional importance. *Biol. Trace Elem. Res.* 32:3-8. 1992.

20.   Mujika, L. & Padilla, S. (1997). Creatine supplementation as an ergogenic aid for sports performance in highly trained athletes: a critical review. *Int. J. Sports Med.* 18:491-496.

21.   Position paper of the American Dietetic Association. (1995). Food and nutrition misinformation. *J. Am. Diet. Assoc.* 95:705.

22.   Position paper of the American Dietetic Association. (1993). Nutrition for physical fitness and athletic performance for adults. *J. Am. Diet. Assoc.* 93:691.

23.   Position paper of the American Dietetic Association. (1994). Enrichment and fortification of foods and dietary supplements. *J. Am. Diet. Assoc.* 94:661.

24.   Position paper of the American Dietetic Association. (1996). Vitamin and mineral supplementation. *J. Am. Diet. Assoc.* 96:73.

25.   SCAN's Guide to Nutrition and Fitness Resources: SCAN'S PULSE, supp. (1998). *The Am. Diet. Assoc.* Chicago, IL.

26.   Statement of FANSA[1]: (1997). What does the public need to know about dietary supplements? *J.Am. Diet. Assoc.* 97:728.

27.   Suminski, R.R., Robertson,R.J., Gross, F.L., Arslanian, S., Kang, J., DaSilva, S., Utter, A.C., & Metz, K.F. (1997). Acute effect of amino acids ingestion and resistance exercise on plasma growth hormone concentration in young men. *Int. J. Sport Nutr.* 7:48-60.

28.   Volek, J.S., & Kraemer, W.J. (1996). Creatine supplementation: Its effect on human muscular performance and body composition. *J. Strength Cond. Res.* 10:200-210.

29.   Volek, J.S., Kraemer, W. J., Bush, J.A., Boetes, M., Incledon, T., Clark, K. L., & Lynch, J.M. (1997). Creatine supplementation enhances muscular performance during high-intensity resistance exercise. *J. Am. Diet. Assoc.* 97:765-770.

30.   Walberg-Rankin, J., Hawkins, C.E., Fild, D.S., & Sebolt, D.R. (1994). The effect of oral arginine during energy restriction in male weight trainers. *J. Strength Cond. Res.* 8:170-77.

31.   Whitney, E.N., Cataldo, C.B., & Rolfes, S.R. (1998). *Understanding Normal and Clinical Nutrition*, 5th Ed. West/Wadsworth.

32.   Ziegenfuss, T.M., Lemon, P.W.R., Rogers, M.R., Ross, R., & Yarasheki, K.E. (1997). Acute creatine ingestion: Effects on muscle volume, anaerobic power, fluid volumes, and protein turnover. *Med. Sci. Sports Exerc.* 29:S127.

[1] The Food and Nutrition Science Alliance, a partnership of four professional scientific societies: American Dietetic Association, American Society for Clinical Nutrition, American Society for Nutritional Sciences, and the Institute of Food Technologists.

CASE STUDY #3
NUTRITIONAL SCREENING

INTRODUCTION

This study provides the student with the opportunity to screen a patient by means of medical history, nutritional history, biochemical data, anthropometric data, and clinical data. An interview is included but, although it was conducted by a registered dietitian, this does not mean that it was done correctly. The purpose of each of the case studies is to gradually introduce the student to the process of reading and understanding a patient's chart, interpreting pertinent data, and assimilating that data into an appropriate nutritional care plan. This case study introduces the student to hospital procedures, particularly nutritional screening. Abbreviations, medical terms, and lab values are more important in this case study than they were in the previous studies.

SKILLS NEEDED

ABBREVIATIONS:

Knowledge of the following abbreviations is required in order to understand this case. You should learn these abbreviations before you begin to read the study.

AIDS : acquired immune deficiency syndrome
ASAP : as soon as possible
BMI : body mass index
CHF : congestive heart failure
Cl : chloride
C/O : complains of
cm : centimeter
CVA : cerebrovascular accident (stroke)
dl : deciliter or 1/10 of a liter
Dx : diagnosis
FH : family history
g : gram
GI : gastrointestinal
Hct : hematocrit
Hgb : hemoglobin
HIV+ : positive carrier for the AIDS virus
Ht : height
HTN : hypertension
ICU : intensive care unit
K : potassium
L : liter
Lymph : lymphocytes
MAC : midarm circumference
MAMC : midarm muscle circumference
MCH : mean corpuscular hemoglobin
MCV : mean corpuscular volume

MD : medical doctor
mEq : milliequivalent
mg : milligram
mg/dl : milligram per deciliter
MH : medical history
mm : millimeter
Na : sodium
NH : nutritional history
N/V : nausea/vomiting
P : phosphorous
pg : picogram
RBW : reference body weight
RD : registered dietitian
Rm : room
SBR : strict bed rest
Ser Alb : serum albumin
SH : social history
TLC : total lymphocyte count
TF : tube feeding
TPN : total parental nutrition
TSF : triceps skin folds
UBW : usual body weight
WBC : white blood cell count
Wt : weight
YOWF : year old white female
m^3 : cubic microns
: number

LABORATORY VALUES:

You will need to be able to interpret the nutritional significance of the following laboratory values for this case study: Cl, glucose, Hct, Hgb, K, lymphocytes, MCH, MCV, Na, Ser Alb, P, and WBC (see Appendix B).

FORMULAS:

The formulas used in this case study include reference body weight using the Hamwi formula, percent reference body weight, percent usual body weight, midarm muscle circumference, total lymphocyte count, basal energy expenditure using the Harris-Benedict equation, total energy expenditure using an appropriate activity factor, and effects of fever on energy needs (most of which can be found in Appendix A; activity factors can be found in Appendix D).

MK is a 21 YOWF college student who has been bothered by a cold for most of the fall semester. She has several colds a year but they usually do not last as long as this one has. She had the usual symptoms: a sore throat, blocked sinuses, and a hacking cough with moderate quantities of slightly yellowish sputum. Two days ago her cough was considerably worse. The amount of sputum increased and was purulent. MK also had a temperature of 99.8° and C/O of pain in the left side of her chest on breathing. Her respirations were shallow. She felt weak and ached all over. She felt so bad that she did not feel like eating anything. Yesterday she was getting ready to take a final exam when she collapsed in her room. Her roommate and some friends brought her to the campus infirmary. The MD listened to her breath sounds and diagnosed pneumonia with exhaustion. She was admitted to a local hospital. Her admission orders included SBR, appropriate antibiotics, respiratory treatments, and a regular diet.

Normally, dietitians do not visit patients with an order for a regular diet unless they are requested to do so. However, this hospital has an excellent nutritional screening process that requires a diet technician to review admission criteria, height, weight, and admission lab values. All of the above information goes into the hospital computer on admission. The technician accomplishes the screening by simply pulling up the new admissions for the previous day on the computer screen and reviewing the four parameters previously mentioned. This can be done while sitting in the dietary offices. The techs have a form to fill out as they review the computerized chart. The form will vary from hospital to hospital, but there will be a thread of commonality in all screening forms. If the form requires extensive information, not all of the information required may be obtainable from the computer screen. There are times when the tech will have to visit the patient to obtain additional information. The day after admission, MK was screened using the computer and a brief interview by the diet tech. The following information was obtained:

MK's MH was not remarkable except for frequent colds. Her weight history was very significant. She claimed that her UBW was 115 lbs and that she had lost 17 lbs over the last three months. She attributed this weight loss to a very strenuous semester. The computer indicated her measured height and weight to be:

▶ **Height 5'4"**

▶ **Weight 98 lbs**

Fasting blood was drawn the morning after admission and the results were as follows:

TEST	RESULT	NORM	TEST	RESULT	NORM	TEST	RESULT	NORM
Hgb	11.0 g/dl	12 - 16	MCH	24 pg	26 - 34	MCV	80 µm³	82 - 98
Hct	34%	36 - 48	WBC	14.5x10³ /mm³	5 - 10 x 10³	Lymph	9.0%	20 - 40
Ser Alb	3.4 g/dl	3.9 - 5.0	Na	135 mEq/L	135 - 145	K	3.0 mEq/L	3.5 - 5.3
Glucose	70 mg/dl	65 - 110	Cl	103 mEq/L	98 - 106	P	2.6 mg/dl	2.5 - 4.5

In order to complete the screening form, the tech had to complete some calculations. Answer the following questions and compare your answers to those found in the following screening form.

1.   Determine MK's RBW (Appendix A, Table A - 7).

$$128$$

2.   Calculate the percent of her RBW, and the percent of her UBW. Please show work for all calculations completed in this report. (Appendix A, Tables A - 9 and 10)

$$\frac{98}{128} \times 100 = 76.56\%$$

$$\frac{98}{115} \times 100 = 85.21\%$$

3.   Calculate MK's BMI and explain what the results mean (Appendix A, Tables 12 and 13).

$$\frac{98}{5.42} = 18.14$$

4.   Calculate her TLC (Appendix A, Table A - 15).

$$(14.5 \times 10^3) \times (9.0) = 130.5 \times 10^3$$

As can be seen on the following screening form, MK's results indicated a high nutritional risk. The diet tech brought the results directly to the dietitian. The dietitian went to see MK to obtain a nutritional history and to do a complete nutritional assessment. Since MK's first day in the hospital was not typical, the dietitian asked MK to describe her intake and activity on a typical day. The dietitian also obtained a food frequency and a family, medical, and social history. MK's MH was as previously mentioned. Her FH was not remarkable. Her SH was another story. She

COUNTY HOSPITAL
NUTRITION SERVICES

NUTRITION SCREENING

Name MK **MD** Dr. JK **Rm #** 1102

Admission Date 3/21/95 **Dx** Pneumonia with exhaustion **Age** 21

Diet Order Regular diet **Ht** 5'4" **Wt** 98 lbs **UBW** 115 lbs **RBW** 128

BMI_____ **Criteria for Nutritional Risk:**

	MILD RISK		MODERATE RISK		SEVERE RISK
	Unintentional weight loss of 5 to 10 lbs in 6 mos		Unintentional weight loss of 10 to 15 lbs in 6 mos	X	Unintentional weight loss of > 15 lbs in 6 mos
	> 20% over RBW		Morbid obesity		> 200% over RBW
	BMI = 19 or BMI = 25 to 30		BMI = 17-18 or BMI = 30 - 35	X	BMI < 17 or BMI > 35
	Mild N/V, diarrhea		Prolonged N/V, diarrhea		Malabsorption
X	Decreased appetite		Very poor appetite		TPN or TF dependency
	Chewing or swallowing problems, other mild nutritional problems		Mild decubitus or other open wounds		Severe decubiti or other wounds that are not healing
	HTN		Renal disease		Severe pancreatic disease
	Atherosclerosis, elevated lipid profile		Early stages of cancer and/or chemotherapy and/or radiation therapy/HIV +		Advanced cancer with cachexia AIDS
	Recent minor surgery/hospitalization		Recent major surgery/hospitalization		GI surgery or major surgery
X	Anemia		Diabetes in poor control	?	Malnutrition
	Ulcer		GI diseases or GI bleeding, ileus, etc.		ICU patient, burns
	Confined or nursing home pt		CHF		Sepsis
?	Simple dehydration		CVA		Multiple trauma
X	Albumin 3.2 - 3.4 mg/dl		Albumin 2.8 - 3.1 mg/dl		Albumin < 2.8 mg/dl
X	TLC 1200 - 1500 cells/mm^3		TLC 900 - 1200 cells/mm^3		TLC < 900 cells/mm^3
	Mild depression		Moderate depression		Severe depression
X	Other:Mild temp/cold		Other:		Other:

High Risk = 1 or more severe risk factors or 3 or more moderate risk factors or 6 or more mild risk factors
Moderate Risk = 2 or more moderate risk factors or 4-6 mild risk factors
Low Risk = < 4 mild risk factors

High Risk = RD to see patient ASAP, do complete assessment, chart, re-evaluate in 3-5 days.
Moderate Risk = RD to see patient in within 3 days, assessment and chart as necessary, re-evaluate in 3-5 days.
Low risk = Basic nutrition services, diet tech to re-evaluate within 7-10 days.

comes from a well-to-do family that expects a lot from her. MK is active in many organizations on campus, including a sorority, cheerleading, student government, and others. She is an honor student and a perfectionist. She claims to drink when she goes out because it is the "in thing" to do. MK denies the use of drugs. She admits to smoking when she is by herself but not in public. Her current class work is the hardest yet and she holds an office in three major organizations on campus. She strongly denied anorexia nervosa or bulimia, but she did admit that she watches her weight very closely. Between cheerleader practice, aerobics, swimming, and intramural athletics, her exercise was classified as heavy. She also admitted to taking laxatives "rather often" for constipation but would not admit to an amount or frequency.

The dietitian's interview of MK's "typical day recall" went like this:

RD: MK, I want you to take me through a typical day at school, when you are feeling well. When you get up in the morning, what is the first thing you have to eat or drink?

MK: I don't eat breakfast on a usual day. I do not have time. I usually have some coffee later in the morning. I need coffee to keep me going.

RD: What do you put in your coffee?

MK: Nothing. I never use sugar or cream . . . fattening

RD: How much coffee do you have in a typical day?

MK: About six or seven cups, mostly at night to help me stay awake to study.

RD: Is this decaffeinated or regular coffee?

MK: Regular! I drink it for the caffeine.

RD: Well, after your morning coffee, what is the next thing you have to eat or drink?

MK: Sometimes I'll pick up a pack of peanut butter or cheese crackers with the coffee if I have time. Otherwise it will be lunch before I'll eat.

RD: And what is a typical lunch like?

MK: Oh, a hamburger or hot dog with some fries and a diet soda. Sometimes it's a pack of crackers again with a diet soda, depending on how much time I have.

RD: Do you put mayonnaise on your hamburger?

MK: No, mustard . . . less calories.

RD: Do you ever eat in the cafeteria on campus?

MK: Are you kidding? My daddy buys me a meal ticket every semester, but I only eat there in the evenings . . . sometimes. The lines are too long, and I don't like their food . . . too much fat in it... and the "mystery meat," ugh!

RD: OK, I understand. When do you eat again?

MK: Well, I have cheerleader practice at 5 P.M. and usually grab a diet soda and a candy bar before practice. I don't like eating sweets, but I have found that if I don't, I will not make it through practice. Besides, I have one of those good candy bars, you know, with nuts in it for energy.

RD: These diet sodas you are drinking, what kind are they? Do they contain caffeine?

MK: Yep! The highest in caffeine I can find.

RD: Do you eat after practice?

MK: Yeah. By the time I get cleaned up, it is time for the cafeteria to close and the lines are not too long. I usually have a salad.

RD: Describe the salad for me.

MK: Well, you know, whatever they have, lettuce, tomatoes, celery, stuff like that.

RD: Do you put anything on your salad?

MK: You mean like dressing? Yea. Usually low-cal Thousand Island.

RD: And do you have anything to drink with this salad?

MK: Yea. I have coffee. I usually have a meeting to go to after eating, and then I have to study. I have a 3.5 GPA you know.

RD: Do you have anything to eat or drink before you go to bed?

MK: Like I told you, I drink coffee while I'm studying to help me stay awake. Sometimes I snack on some cookies or something.

RD: You did not mention milk. Do you ever drink milk?

MK: Sometimes, if I ever eat breakfast, I use some on my cereal.

RD: What about cheese or ice cream?

MK: I like cheese. I eat it when I'm at home, but I don't have a place to keep it here. I eat ice cream occasionally. It's too fattening. Sometimes I get some low-fat frozen yogurt.

RD: You did not mention fruit. Do you ever eat fruit?

MK: I like fruit, but I don't buy it here; I do at home. If I eat breakfast, I drink orange juice.

RD: What about vegetables? Other than the salad and the potatoes, you did not mention vegetables.

MK: I don't like vegetables. I eat some carrots or green beans at home. I only like them the way my mother fixes them.

The dietitian finished her interview and obtained some anthropometric measurements, the results of which were as follows:

► **TSF 18 mm**

► **MAC 25.7 cm**

While the dietitian was talking to MK, she noticed her hair was very dull and stringy for a cheerleader. She was very thin and appeared weak and malnourished. She was pale and, upon examination, her nail beds were pale and slow to return to their normal color when pressed. Her skin was dry and appeared to be "itchy." MK scratched her arms several times while the dietitian was there.
****

QUESTIONS:

5.   Briefly define the following terms:

Sputum: chest culture of the mucous

Bulimia: an eating disorder characterized by repeated episodes of binging followed by self induced vommiting

Purulent: puss

Anthropometry: relating to the measurement of the physical characteristics of the body

Anorexia nervosa: an eating disorder characterized by a refusal to maintain a minimally normal body weight

6.   Calculate MK's daily energy requirements using the Harris-Benedict equation and an activity factor (Appendix D, Table D - 1 through 3 and D - 5).

W = 98 lbs
H = 5'4
A = 21

$$655.1 + (9.6 \times Wkg) + (1.8 \times Hcm) - (4.7 \times A) =$$

44.54   13,716

$$655.1 + 427.584 + 24.688 - 98.7$$

$$1008.672$$

7.   At the onset of MK's sickness, she had a temperature of 99.8°. How many kcals would this require above her hospitalized BEE? (Appendix D, Table D - 4).

① $99.8° - 98.6° = 1.2$

② $1.2 \times 7\% = 8.4 \% BEE$

③ $1008.672 \times 8.4\% = 84.72$

④ $1008 + 84.72 = 1092.72$

8. Convert 99.8° F to °C (Appendix A, Table A - 4).

$$\frac{5}{9}(t_f - 32) = t_c$$
67.8

37.6°C

9. Calculate MK's midarm muscle circumference (Appendix A, Table A - 14).

MAC (cm) − (3.14 × Skinfold (mm)) =
25.7cm − (3.14 × 18mm) =
25.7 − 56.52 = −
− 30.82

10. What percentile is MK in for TSF, MAC, and MAMC and what do these mean?

TSF = 15%

11. Analyze MK's nutritional intake and name the possible vitamin and mineral deficiencies that could exist.

Iron deficient

12. List all abnormal lab values that have nutrition implications and tell what those implications are.

13. Explain how MK's lifestyle and/or diet may have affected the lab values you mentioned above.

14. Does MK have kwashiorkor, marasmus, both, or neither? Explain (see Weinsier, R.L, Heimburger, D.C. & Butterworth, C.E. *Handbook of Clinical Nutrition,* 2nd Ed. 1989; pg. 134 for an explanation of this).

> She has neither. Some of the effects of chronic malnutrition are decreased activity. MK does not decreased activity. But on the other hand she has pneumonitis and that gives evidence of malnutrition

15. After the interview was over and the anthropometric measurements were taken, the RD made some observations concerning MK's hair, general appearance, color, nail beds, and skin. What are these observations called? What does each indicate?

> These measurements are what RD's use for perception.

16. The RD asked some questions at the end of her interview in an attempt to obtain the intake of certain specific nutrients. Which specific nutrients should any registered dietitian be concerned about regardless of the disease state of the client being interviewed?

17. Read the case again very carefully. Think about all the possible causes of stress in MK's life. List these causes, including mental, physical, and social.

Cheerleading practice at 5:00. She works really hard on maintaining her 3.4 GPA, and has meetings right after cheerleading practice.

18. As a health care professional, you will counsel patients like MK. You will not be expected to be their psychologist, but you will be expected to be sensitive to their needs and be able to offer some alternatives to their lifestyle. Unless MK makes some changes, she is not going to have time to eat and will never be able to correct her nutritional deficiencies. What are some suggestions you could give MK about the stressful situations you mentioned above?

You should make time in the morning to bag yourself a lunch + snack for the duration of the day.

19. There are a few techniques the RD used in the interview that were wrong. See if you can find the errors and indicate how you would correct them.

She kept saying you did not mention something, like she was suposed to mention it.

20. List at least three good points about this interview.

She was very specific about certain foods.

21. MK strongly denied anorexia nervosa and bulimia, but there were several indications in her interview that suggested either anorexia or bulimia. List these indications.

Using coffee alot, or drinking alot of caffenine. Cafenine is used as a dieretic

22.   Caffeine was high in MK's intake. What are the possible effects of caffeine on the human body?

Dehydrate you, and cause you to go to the bathroom

23.   Would the laxatives have an effect on any of the lab values? If so, which ones and how?

24.   Briefly discuss MK's opinion of the candy bar with nuts.

the candy bar with nuts contain energy.

25.   Briefly outline how you would counsel MK on her diet. Include in your outline behavior modification tips and the following topics:
A) The nutrients she is lacking and how she could increase those nutrients through diet.
B) The substances she has an excess of and how she could avoid those substances.
C) How she can incorporate these principles into her lifestyle.
D) How you would try to convince MK of the importance of these changes.

RELATED READINGS

1.   Charney, P. (1995). Nutrition assessment in the 1990s: where are we now? *Nutr. Clin. Pract.* 10(4):131-139.

2.   Davis, C., Kaptein, S., Kaplan, A.S., Olmsted, M.P., & Woodside, D.B. (1998). Obsessionality in anorexia nervosa: the moderating influence of exercise. *Psychosom. Med.* 60(2):192-197.

3.   Elmore, M.F., Wagner, D.R., Knoll, D.M., Eizember, L., Oswalt, M.A., Glowinski, E.A., & Rapp, P.A. (1994). Developing an effective adult nutrition screening tool for a community hospital. *J. Am. Diet. Assoc.* 94(10):1113-1118.

4.   Food and Nutrition Board, National Research Council. (1989). *Recommend Dietary Allowances,*10th Ed. Washington: National Academy Press, pp. 52-77.

5.   Grindel, C.G. & Costello, M.C. (1996). Nutrition screening: an essential assessment parameter. *Medsurg. Nurs.* 5(3):145-154.

6.   Kerekes, J. & Thornton, O. (1996). Incorporating nutritional risk screening with case management initiatives. *Nutr. Clin. Pract.* 11(3):95-97.

7.   Kovacevich, D.S., Boney, A.R., Braunschweig, C.L., Perez, A, & Stevens, M. (1997). Nutrition risk classification: a reproducible and valid tool for nurses. *Nutr. Clin. Pract.* 12(1):20-25.

8.   Lee, R.D. & D.C. Nieman. (1993). *Nutritional Assessment*. Brown and Benchmark.

9.   McLaren, S. & Green, S. (1998). Nutritional screening and assessment. *Prof. Nurse.* 13(6 S):S9-S15.

10.   Nagel, M.R. (1993). Nutrition screening: identifying patients at risk for malnutrition. *Nutr. Clin. Pract.* 8(4):171-175.

11.   Position paper of the American Dietetic Association. (1994). Nutrition intervention in the treatment of anorexia nervosa, bulimia nervosa, and binge eating. *J. Am. Diet. Assoc.* 94:902.

12.   Position paper of the American Dietetic Association. (1997). Weight management. *J. Am. Diet. Assoc.* 97:71.

13.   Pryor, T., & Wiederman, M.W. (1998). Personality features and expressed concerns of adolescents with eating disorders. *Adolescence.* 33(130):291-300.

14.   Siegel, M., Brisman, J., & Weinshel, M. (1989). *Surviving an Eating Disorder.* New York: Harper & Row.

15.   Sullivan, P.F., Bulik, C.M., Carter, F.A., Gendall, K.A., & Joyce, P.R. (1996). The significance of a prior history of anorexia in bulimia nervosa. *Int. J. Eat. Disord.* 20(3):253-261.

16. Weinsier, R.L., Heimburger, D.C., & Butterworth, C.E. (1989) *Handbook of Clinical Nutrition,* 2nd Ed. St. Louis. C.V. Mosby Company. pg. 134.

17. Whitney, E.N., Cataldo, C.B., & Rolfes, S.R. (1998). *Understanding Normal and Clinical Nutrition,* 5th Ed. West/Wadsworth.

18. Wills-Brandon, C. (1989). *Eat Like a Lady. Guide for Overcoming Bulimia.* Deerfield Beach, Flordia. Health Communications, Inc.

CASE STUDY #4
NUTRITIONAL ASSESSMENT

INTRODUCTION
This study is concerned with nutritional assessment, general problems facing the elderly, anemia, dehydration, and normal laboratory values. It is also a good introduction to the use of some medications. If the medications are too advanced, omit questions 7 - 11.

SKILLS NEEDED

ABBREVIATIONS:
Knowledge of the following abbreviations is required in order to understand this case. You should learn these abbreviations before you begin to read the study.

Alk Phos : alkaline phosphatase
ALT : alanine aminotransferase
amp : ampule
ASA : aspirin
AST : aspartate aminotransferase
Bili : bilirubin
c : with
BR : bed rest
BUN : blood urea nitrogen
Ca : calcium
CC : chief complaint
cc : cubic centimeter
Cl : chloride
CPK : creatine phosphokinase
Cr : creatine
DBW : desirable body weight
d : day
dl : deciliter
Dx : diagnosis
ER : emergency room
ETOH : alcohol
FF : force fluids
Fx : fracture
g : gram
GI : gastrointestinal
glu : glucose
HA : headache
Hct : hematocrit
Hgb : hemoglobin

Mg : magnesium
$MgSO_4$: magnesium sulfate
MH : medical history
mm^3 : cubic millimeter
MNT : medical nutrition therapy
MOM : milk of magnesia
M.T.E. - 4 : four mineral trace elements
MVI - 12 : multiple vitamin infusion
Na : sodium
N/V : nausea and vomiting
P : phosphorous
po : by mouth
pt : patient
qd : every day
Ⓡ : right
RBW : reference body weight
RDA : recommended dietary allowance
RD : registered dietitian
R/O : rule out
Rx : prescription
Ser Alb : serum albumin
t.i.d. : three times a day
TLC : total lymphocyte count
UBW : usual body weight
U/L : units per liter
WBC : white blood cell count
x : times
YO : year old
YOM : year old male

H&H : hematocrit and hemoglobin
Hx : history
IM : intramuscular
I.V. : intravenous
K : potassium
L : left
Ⓛ : liter
lymph : lymphocytes
MCV : mean corpuscular volume
mEq : milliequivalent
mg : milligram

YOWM : year old white male
3d : 3 days
μm^3 : cubic microns
♂ : male
@ : at
↓ : less or decreases
↑ : more or increases
∅ : none or zero
D_5NS : dextrose 5% in water
Ⅰ : one
p̄ : after

LABORATORY VALUES:

You will need to be able to interpret the nutritional significance of the following laboratory values for this case study: Alk Phos, Bili, BUN, Ca, Cl, CPK, Cr, Glucose, Hct, Hgb, K, Lymphocytes, MCV, Mg, Na, P, Ser Alb, ALT, AST, Serum Amylase, and WBC (see Appendix B).

FORMULAS:

The formulas used in this case study include reference body weight using the Hamwi formula, percent reference body weight, total lymphocyte count, basal energy expenditure using the Harris-Benedict equation, and an appropriate stress factor (See Appendices A and D).

MEDICATIONS:

Become familiar with the following medications before reading this case study. Note the diet-drug interactions, dosages and method of administration, gastrointestinal tract reactions, etc.

1. Milk of Magnesia; 2. Haldol (haloperidol); 3. Norpace (disopyramide phosphate); 4. Pepcid (famotidine); 5. Clinoril (sulindac); 6. Dalmane (flurazepam hydrochloride); 7. Di-Gel Liquid; 8. Aspirin (acetylsalicylic acid).

Mr. D is a 73 YOWM who has lived with his son for the past five years. He ambulates well but his mental powers are slipping. He has a hard time remembering from one day to the next. His son has noticed his father's condition deteriorating significantly over the last six months. Because of his condition and the family's inability to provide proper attention at home, he should have been placed in a nursing home, but he refused to go. His son knows it is best for his father but he does not have the heart to admit him.

Mr. D had not been eating well. He had been losing weight and growing weaker. He said he just did not feel like eating. Nothing tasted the same. He has dentures but refused to wear them. He claimed they were too loose and would fall out if he tried to eat with them. His daughter-in-law does the cooking, but she does not always have the time to fix soft foods especially for him. She tried feeding him pureed food, but he absolutely refused to eat "baby food." His eyesight and hearing have also been failing. He was being treated for many disorders, none of which were serious, but he was taking several medications. This created another problem in light of his failing memory. He either forgot to take his medicine, or he took too much. His medications included: haloperidol (Haldol); disopyramide phosphate (Norpace); Milk of Magnesia; famotidine (Pepcid); sulindac (Clinoril); flurazepam hydrochloride (Dalmane); Di-Gel Liquid; and aspirin for frequent headaches. One day his son and daughter-in-law returned home from work and found him on the floor, unable to get up. He was in pain and was not able to move his left leg. His son called for an ambulance and brought him to the hospital.

His Dxs were:
- Fx left femur
- Cachexia
- Dehydration
- R/O malignancy

Normally, weighing a patient with a broken leg would not be attempted, but because Mr. D was so thin, a bed weight was obtained after stabilization of the Fx. Mr. D weighed 133 lbs and was 6'0" tall.

His son was embarrassed by the Dx and his father's weight. He said: "I knew my father was losing weight. I did not realize he lost this much." Mr. D weighed 165 lbs a year ago.

Lab values on admission:

TEST	RESULT	NORM	TEST	RESULT	NORM	TEST	RESULT	NORM
Hgb	12.8 g/dl	14 - 17.4	Mg	1.3 mEq/L	1.3 - 2.1	MCV	80.8 μm^3	82 - 98
Hct	38%	42 - 52	WBC	4.3x10^3 /mm^3	5 - 10	Lymph	18.0%	20 - 40
Ser Alb	3.8 g/dl	3.9 - 5.0	Na	135 mEq/L	135 - 145	K	4.2 mEq/L	3.5 - 5.3
Glucose	181 mg/dl	65 110	Cl	106 mEq/L	98 - 106	P	3.8 mg/dl	2.5 - 4.5
BUN	40 mg/dl	7 - 18	Cr	1.1 mg/dl	0.6 - 1.3	Ca	10.8 mg/dl	8.6 - 10
CPK	325 U/L	38 - 174	AST	18 U/L	5 - 40	ALT	12 U/L	7 - 56
Serum Amylase	62 U/L	25 - 125	Alk Phos	53 U/L	17 - 142	Bili	1.6 mg/dl	0.2 - 1.0

Mr. D's Medical Nutrition Therapy (MNT) included:
- High protein mechanical soft (edentulous) diet with snacks
- Force fluids
- Have dietitian see pt.
- Thiamin, 50 mg IM each hip qd x 3d
- MgSO$_4$ 1g right hip qd x 3d
- Folate 1 mg po qd
- D$_5$W @ 77 cc/h
- 1 amp of MVI-12 and 9 cc M.T.E.- 4 (see Appendix E) in one I.V. bottle qd

Lab values several days later after I.V. hydration:

TEST	RESULT	NORM	TEST	RESULT	NORM	TEST	RESULT	NORM
Hgb	10.6 g/dl	14 - 17.4	Mg	1.9 mEq/L	1.3 - 2.1	MCV	79 μm^3	82 - 98
Hct	35.3%	42 - 52	WBC	5.2×10^3 /mm^3	5 - 10	Lymph	19.0%	20 - 40
Ser Alb	2.8 g/dl	3.9 - 5.0	Na	134 mEq/L	135 - 145	K	4.4 mEq/L	3.5 - 5.3
Glucose	127 mg/dl	65 - 110	Cl	104 mEq/L	98 - 106	P	4.2 mg/dl	2.5 - 4.5
BUN	16 mg/dl	7 - 18	Cr	0.6 mg/dl	0.6 - 1.3	Ca	8.0 mg/dl	8.6 - 10
CPK	150 U/L	38 - 174	AST	21 U/L	5 - 40	ALT	20 U/L	7 - 56
Serum Amylase	56 U/L	25 - 125	Alk Phos	269 U/L	17 - 142	Bili	0.9 mg/dl	0.2 - 1.0

The RD went to Mr. D's nursing station and read his chart. In so doing, she was able to verify the physician's orders and his synopsis of the patient. After reading the chart, the RD visited Mr. D but, because of his mental status, did not try to interview him. She did interview his son to obtain his father's likes and dislikes. Anthropometric measurements were not obtained since it had already been determined that Mr. D was a nutritional risk. His anticipated hospital stay was not long. The son was now planning to place his father in a nursing home so follow-up was not likely. After visiting the patient and collecting as much information as she could, the RD left a chart note to document her visit. In her chart note, she indicated her observations and stated that she, or her diet tech, would visit the patient on a daily basis during the noon meal to evaluate meal acceptance. She also indicated that she would initiate a calorie count to determine actual daily energy and protein intakes. She did not recommend further tests, as for anergy, since malnutrition had already been established and was being treated.

Chart notes are often difficult to read because they are written in haste. What follows are examples of a physician's and a dietitian's progress note. The legibility of the notes is good when compared to many in real life. The patient's chart will also contain a Doctor's Order sheet on which he will record his orders. This sheet is not included here but the physician repeats his orders in his progress note. This is the procedure that many physicians often follow. Locate each of the orders mentioned above and note how they are written. Attempt to read the following notes as practice for the real world. There is always more than one way to write a chart note. The dietitian's note should describe the patient's problem (or problems) from a medical, physical, social, and nutritional point of view. The information box below describes how to accomplish this. See if the chart note complies with the information box. As you read the notes, think about what changes you would make, if any, to make the note more readable or informative.
****

Information Box 4 - 1

Following is a mock progress note that contains a physician's comments and the RD's note. The RD has used a widely accepted style called a "SOAP" note:

Subjective: In the first part of the note, subjective information is recorded. This is any information obtained verbally from the patient or anyone associated with the patient, such as a family member or visitor who knows the patient. Examples of the information in this section include:
- Nutritional history, food likes and dislikes, current intake, appetite.
- Usual body weight.
- Usual activity, type of work, exercise.
- Socioeconomic status, cultural habits, pertinent information about family.
- Physical impairments, dentures.

Objective: The second section records the objective information. This includes all information that can be measured or documented in some way:
- All anthropometric information, as height, weight, and skinfolds.
- Reference body weight or ideal body weight.
- Recorded information in the chart, as diet order, age, diagnosis.
- Pertinent results of diagnostic tests, laboratory results.
- Nutritionally important results of surgical procedures.
- Nutritionally related medications.
- Results of calorie counts, computerized diet analysis.

Assessment: The third section includes the RD's assessment based on the information obtained in **S** and **O**:
- Evaluation of nutritional status based on history, appetite, intake.
- Evaluation based on anthropometrics, usual body weight.
- Interpretation of lab values and diagnostic tests as they pertain to nutritional deficiencies or requirements.
- Estimation of nutrient deficiencies and requirements based on above information.
- Assessment of appropriateness of diet order to meet the patient's needs.
- Assessment of the patient's comprehension, ability, and enthusiasm about following diet.
- Estimation of the patient's level of compliance.

Plan: The last section is the plan and relates what the RD is going to do about the assessment:
- Goals for nutritional therapy.
- Recommendation for changes in diet order and/or additional supplements, or approval of the current regimen, as appropriate.
- Plans for follow-up, future visits, encouragement at mealtime.
- Suggestions for additional tests necessary for a more complete assessment, if appropriate.
- Suggestions for referrals to a social worker, physical therapist, etc., as appropriate.
- Description of nutritional education given, to be given, or requested to be given, as appropriate.

PROGRESS NOTES

8/1/98

CC: This malnourished cachexic 73 YOW ♂ admitted via E.R. c̄ fx ℗ femur, malnutrition & dementia. Pt found @ home on floor conscious but in pain & unable to move ℗ leg. X-ray confirms ℗ femur fx severely malnourished. Labs pending — suspect they will confirm dehydration.

MH: Hx of GI distress - gastric ulcers, constipation. Frequent HA resulting in ASA abuse. Clinoril for arthritis. Dentures, ↓ eyesight & ↓ hearing. Ø ETOH abuse/tobacco abuse.

℞: BR, FF to hydrate, ↑ prot mech soft diet c̄ nourishments t.i.d. D₅NS c̄ 75 cc/h c̄ ↑ amp MVI ↑ MTE - 4 ✓ d. Thiamine 50 mg ↑ M ea hip ✓d x 3d, M₃ SO₄ 1 g ℗ hip ✓d x 3d, folate 1 mg p o ✓ d. R/O malignancy / Alzheimer's. R.D. to see pt.

8/1/98
15 30

Nutrition: S̲: Pt's son states: "My father has been depressed and has not been eating well. He has lost wt. gradually over the past yr." Food preferences were obtained from the son.

O: 73 YOW ♂ Ht. 6' Wt 133 UBW 165 RBW 178 ± 10%. Labs: ↑ glu 181 mg/dl, ↑ BUN 40 mg/dl, ↑ CPK 325 U/L; N↑N. Diet order: High protein mechanical soft diet c̄ nourishments. Dx: Malnutrition, fx ℗ femur, dementia, dehydration.

A: 32# wt loss in past year. 75% of RBW & 81% UBW. Pt dehydrated — when rehydrated, expect to see labs change & provide more indications of malnutrition. agree c̄ nutritional plan.

P: Will send diet as ordered c̄ ↑ kcal ↑ prot nourishments t.i.d. Will visit pt. c̄ diet tech during meal time to evaluate intake and will start a calorie count X 3 days to determine ↑ cal & prot. intake. Will reevaluate p̄ 3 days of kcal. count & pt. has been hydrated.

R.D.

QUESTIONS:

1.   Briefly define the following terms:

   Cachexia:

   Edentulous:

   Anergy:

   Anorexia:

   Force fluids:

   Medical Nutrition Therapy:

   Alzheimer's Disease:

2.   Determine Mr. D's RBW and his percent of RBW. Please show work for all calculations (Appendix A, Tables A - 7 through 10).

3.   Calculate Mr. D's BMI and interpret the results (Appendix A, Tables A - 12 and 13).

4.   List the lab values affected by hydration and explain why.

5. Calculate Mr. D's TLC (Appendix A, Table A - 15).

6. List each lab value that suggests a nutritional deficiency. Identify the nutritional deficiency in each case and explain how the circumstances in Mr. D's history contributed to each deficiency.

7. Why should Mr. D be cautioned about taking too many aspirins?

8. Taking multiple medications is a problem that is common with the elderly. Not only does this create possible diet-drug interactions, but it is expensive. Considering many elderly live on fixed incomes, they may be left with insufficient funds for the food they need. Find out if there are any programs in your area that provide assistance for these problems and discuss.

9.   Diet-drug interactions are important in any nutritional assessment. The medications a person has to take and the effects these medications have on nutrient availability must be considered and planned for. Therefore, it is necessary to be familiar with the more common medications. Look up each of the following drugs mentioned in the case study and identify their action. Using the table below, identify those that could have the following complications.

DRUG	ACTION	N/V	CONSTIPATION	DIARRHEA	ANOREXIA
Haldol					
Norpace					
MOM					
Pepcid					
Clinoril					
Dalmane					
Di-Gel					
Aspirin					

10. Identify the medications that are sources of nutrients and indicate the nutrients.

11. Which medications could cause gastric bleeding? Briefly explain how.

12. What are possible causes of Mr. D's lack of taste?

13. Do you agree or disagree with the RD's decision not to take anthropometric measure-ments or recommend anergy determinations? Discuss why or why not.

14. When hospitalized, Mr. D was given 50 mg of thiamin IM qd x 3d. What is the current RDA for thiamin for a 73 YOM? How could such a dose be justified, especially since he is to receive one amp of MVI-12 qd?

15. Look up and record the RDAs for folate and Mg for a 73 YOM. Discuss how you could justify such a dose of folate and Mg. Mr. D is also receiving 9 cc of M.T.E.- 4 (see Appendix E).

16. Compare the RDAs for a 51 YO, a 73 YO, and a 90 YO. Considering the results of these comparisons, what do you think about the sufficiency of the RDAs?

17.   Calculate Mr. D's basal energy expenditure using the Harris-Benedict equation. Determine a stress factor to multiply this by and determine total energy needs (Appendix D, Tables D - 1 and D - 5).

18.   Considering Mr. D's mental and physical condition, his nutritional deficiencies, his medications, and his calculated energy needs, plan a 3-day sample menu for Mr. D that would meet his needs.

19.   Discuss how you might influence Mr. D to eat.

20.   Does Mr. D have symptoms of kwashiorkor, marasmus, both, or neither? Explain.

21.   Suggest a commercial nutritional supplement that would be appropriate here and explain your choice (Appendix F).

22.   Comment on the dietitian's chart note. Was it adequate? Were the components in the right place? Did it need additional information?

▸ **ADDITIONAL OPTIONAL QUESTIONS** ◂

Tube Feeding Drill:

23. Using the table below, compare several of the enteral nutritional supplements that provide about 1 kcal/ml and can be taken orally or with a feeding tube (there is room for seven comparisons).

Product	Producer	Form	Cal/ml	Non-pro Cal/g N	g/L			Na mg	K mg	mOsm /kg Water	Vol to meet RDAs in ml	g of fiber /L	Free water /L in ml
					Pro	CHO	Fat						

24. Using the table below, compare several of the enteral nutritional supplements that provide 1.5 kcal/ml and can be taken orally or with a feeding tube (there is room for seven comparisons).

Product	Producer	Form	Cal/ml	Non-pro Cal/g N	g/L			Na mg	K mg	mOs m/kg Water	Vol to meet RDAs in ml	g of fiber /L	Free water /L in ml
					Pro	CHO	Fat						

Related Readings

1. Chernoff, R. (1991). *Geriatric Nutrition.* 3rd Ed. Gaithersburg, Maryland. Aspen Publishers.

2. Fischbach, F.T. (1995). *A Manual of Laboratory & Diagnostic Tests.* 5th Ed. Philadelphia. J.B. Lippincott Company.

3. Food and Nutrition Board, National Research Council. (1989). *Recommend Dietary Allowances.* 10th Ed. Washington: National Academy Press, pp. 52 - 77.

4. Grant, A. & DeHog, S. (1991). *Nutritional Assessment and Support.* 4th Ed. Seattle, Washington. A. Grant and S. DeHog, Publishers.

5. Lee, R.D. & Nieman, D.C. (1993). Nutritional Assessment. Brown and Benchmark, pp. 137 - 153.

6. Kovacevich, D.S., Boney, A.R., Braunchweig, C.L., Perez, A., & Stevens, M. (1997). Nutritional risk classification: a reproducible and valid tool for nurses. *Nutr. Clin. Pract.* 12(1):20-25.

7. McLaren, S. & Green, S. (1998). Nutritional screening and assessment. *Prof. Nurse.* 13(6S):S9-S15.

8. Nagel, M.R. (1993). Nutrition screening: identifying patient's at risk for malnutrition. *Nutr. Clin. Pract.* 8(4):171-175.

9. Position paper of the American Dietetic Association. (1996). Nutrition, aging, and the continuum of care. *J. Am. Diet. Assoc.* 96:1048.

10. Position paper of the American Dietetic Association. (1998). Liberalized diets for older adults in long-term care. *J. Am. Diet. Assoc.* 98:201.

11. Prendergast, A. & Fulton, F.L. (1997) *Medical terminology: A Text/Workbook.* 4th Ed. Redwood City, California. Addison-Wesley Nursing.

12. Pronsky, Z.M. & Solomon, E. (1997). Food-Medications Interactions. 10th Ed. Phoenix, Arizona. Food-Medications Interactions, Publishers and Distributors.

13. Schlenker, E.D. (1997). *Nutrition in Aging.* St. Louis, Missouri. Mosby-Year Book, Inc.

14. Weinsier, R.L., Heimburger, D.C., & Butterworth, C.E. (1989). Handbook of Clinical Nutrition. 2nd Ed. St. Louis. C.V. Mosby Company.

15. Whitney, E.N., Cataldo, C.B., & Rolfes, S.R. (1998). *Understanding Normal and Clinical Nutrition.* 5th Ed. West/Wadsworth.

CASE STUDY #5
ANOREXIA NERVOSA-BULIMIA NERVOSA

INTRODUCTION
The occurrence of anorexia nervosa is increasing in our society. In the past, the actual incidence of this disorder was not widely known since it has been a disease kept in a "closet." Because of this, the number of cases of anorexia nervosa may be higher than previously realized. The consequences of this eating disorder can be devastating. It is important for health care professionals to understand what anorexia nervosa is, how to recognize it, and what can be done for someone who is suffers from this disorder.

SKILLS NEEDED

ABBREVIATIONS:
Knowledge of the following abbreviations is required in order to understand this case. You should learn these abbreviations before you begin to read the study.

BMI : body mass indes
BMR : basal metabolic rate
kcals : kilocalories

RBW : reference body weight
YOWF : year old white female

FORMULAS:
The formulas used in this case study include reference body weight, percent reference body weight (Appendix A, Tables A - 7 through 10), and basal metabolic rate using the Harris-Benedict equation (Appendix D, Table D - 1), with activity factors for starvation (Appendix D, Table D - 3).

MEDICATIONS:
Become familiar with the following medications before reading the case study. Note the diet-drug interactions, dosages, methods of administration, gastrointestinal tract reactions, etc.

1. Emetics; 2. Diuretics; 3. Laxatives.

SP is a 19 YOWF in her second year of college. She is the only child of an upper-middle class family. She lives at home with her parents, both of whom have flourishing careers and expect her to be highly successful. With all the good intentions in the world, her parents decided to help her reach success by placing tight restraints on her and pressuring her to be the best in everything she attempts. SP has very few opportunities to make decisions for herself. She does not see this as a sign of her parents' love but rather views it as the placement of unreasonable restrictions designed to hinder her social life. SP feels she has to perform to receive love and encouragement, instead of receiving love and encouragement to help her perform. Her parents want her to be a successful doctor but SP wants to be a special education teacher.

SP's parents expect her to be active in as many prestigious campus organizations as she can and still maintain an "A" average. Because of this emphasis on excellence, SP has set very high goals for herself, but her grades in school have been poor because she does not like what she is studying. She cannot put her heart into her studies. In the past, when she failed to reach a goal, she would blame herself and become very depressed. Now she is blaming her parents. At least this is the theory her psychologist has proposed to explain her refusal to eat. He believes she is rebelling against her parents' dominance.

During her freshman year, SP took a health course and learned about anorexia nervosa. After taking the course, SP started talking about how she needed to lose a few pounds and went on a diet. She was 5'2" with a medium frame and weighed 120 lbs. She wanted to go on a diet "the right way," so she obtained a book from a news stand on basic nutrition and calorie content and began counting. She learned that fat provides more calories than carbohydrate or protein, so she tried to eliminate fat from her diet entirely. SP discovered that sugar provides "empty calories," so she tried to eliminate sugar. Her nutrition and calorie counter book said that white flour was harmful, so she eliminated white flour. She also read that a diet high in meat, particularly red meat, was also high in fat and could cause cancer, so she eliminated red meat.

Her nutrition and calorie counter book also emphasized the importance of exercise for weight loss and a healthy body. She started an exercise program that was very vigorous. She attended aerobics classes three times a week, rode her bicycle almost everywhere she went, and played tennis on a regular basis. In addition, as previously mentioned, she was active at school.

SP did not think she was losing weight fast enough so she changed her diet to a semi-starvation diet. She avoided eating with her parents as much as possible to keep them from seeing her starve herself. If she did eat with them, she would eat a normal amount and then go force herself to vomit. When she was down to 105 lbs, her mother noticed a difference and asked her about the weight loss. SP told her she was on a diet. Her mother told her to stop dieting immediately; she had lost enough weight. This encouraged SP to stay on the diet longer. When she reached 100 lbs, she was on a plateau and could not lose additional weight. Her friends were continually telling her how thin she looked. This reinforced her desire to lose weight and gave her a feeling of accomplishment. Her mother nagged her constantly. The more people talked to her about her diet, the more determined she was to lose more weight. She was convinced that she needed to lose a few more pounds. SP decreased her intake even more and increased her exercise. She began to become tired very easily, could not concentrate, and amenorrhea and headaches were a problem. Her grades were getting worse and SP started spending most of her time alone.

One day SP collapsed at school after standing up rapidly. She had to be brought home. Her mother was furious. She took her to a physician who examined SP and said she had orthostatic hypotension and bradycardia. A clinical examination revealed lanugo, and SP admitted having amenorrhea. Her weight was 90 lbs. The doctor easily made a diagnosis of anorexia nervosa. SP denied it and her mother agreed with her at first, refusing to believe her daughter was starving herself. The physician recommended a psychologist and a registered dietitian. Both SP and her mother refused to see them. SP insisted that nothing was wrong with her. Her mother insisted that she was going to handle her daughter her way.

The situation continued until SP became so weak she had to drop out of school. At this point her father demanded she return to the doctor and follow his recommendations. SP started receiving counseling from the psychologist and nutritionist. She weighed 85 lbs.
****

QUESTIONS:

1. Determine SP's RBW and percent RBW (Appendix A, Tables A - 7 through 10). Show all work. $5'2"$ RBW = 120

    %RBW = $\frac{85}{120}$ = 70.8%

2. After her first visit to the physician, SP weighed 90 lbs. Her weight before her diet was 120 lbs. What was her percent loss of weight (Appendix A, Table A - 10)?

    $\begin{array}{r} 120 \\ - 90 \\ \hline 30 \end{array}$ $\frac{30}{120}$ = 25%

3. SP's weight after her last visit was 85 lbs. Recalculate her percent loss of weight.

    $\begin{array}{r} 120 \\ - 85 \\ \hline 35 \end{array}$ $\frac{35}{120}$ = 29%

4. Calculate SP's BMI and comment on its interpretation (Appendix A, Tables A - 12 and 13).

    $\frac{85}{2.2}$ 5.2 × 2.54 = $\frac{wt}{ht^{m2}}$ $\frac{38.63}{(.1320)(.1320)}$ = 2220

    38.63   13.20   0174

    85 × 2.2 / 5.2 ÷ 2.54

5. Define the following terms:

    Emetics: medicine that induces vomiting

    Diuretics: ~~drug~~ induces water loss

    Orthostatic hypotension: blood pressure drop from lying to sitting/standing position of approximately 20 mg -

    Bradycardia: slow heart beat.

Amenorrhea: lack of menstral period

Lanugo:

6. Discuss anorexia nervosa relative to the following points:

♦ What social class(es) of people <u>usually</u> have anorexia nervosa?

♦ What physical, social, mental, or psychological characteristics do individuals with anorexia nervosa frequently have?

① high achievers ④ family overprotectiveness
② low self esteem ⑤ lack family conflict resolution
③ inheritance

♦ What are the theories about the possible causes of anorexia nervosa?

① Obsessive compulsive

♦ What is the estimated incidence of anorexia nervosa?

6-15% is affected between ages 12-13 and 19-20 years of age, other studies say it can occur at any age

7. What are the goals for nutritional therapy for an anorectic like SP?

① Nutritional education
② Correct malnutrition with oral feedings if possible, tube feeding if nessecary
③ Appropriate vitamin/mineral supplementation
④ Nutrition education

8. SP's problem is a very complex one and requires counseling from several different members of the health care team. As a member of that team, you must be careful to reinforce, and not contradict, the information SP is receiving from the other members of the team. Discuss what you think your boundaries are and how the team effort should be coordinated.

9. List the symptoms of anorexia nervosa SP demonstrated. List all additional possible symptoms that SP could have demonstrated.

parents push her for exellent achievments
self distortion of keep wanting to loose weight
over achiever

10. When SP's friends and family told her how thin she looked, it encouraged her. This is typical for anorexics. What approach should be used to discourage rather than encourage a person with anorexia?

Reduce her preoccupation with weight and food, promote adequate self esteem

****

When SP went to see the RD, her mother insisted on going with her. The first interview included the following:

RD: SP, when did you first start on your diet?

SP: Well . . .

Mother: She started a long time ago. Those kids she hangs around with, they talked her into it.

RD: Yes ma'am. When you first started on your diet, how did you see yourself?

SP: I . . .

Mother: She has a very healthy perception of herself; always did. She just wanted to lose a few pounds and then got sick and lost a lot, that's all.

RD: Yes ma'am. SP, I want **you** to tell me everything you have to eat or drink in a typical day from the time you get up in the morning to the time you go to bed at night, everything.

Mother: Now tell the lady what you eat honey, don't leave out anything. She is just trying to help . . . . . .

This interview is not intended for comic relief. Such behavior is not uncommon and has to be dealt with. During the interview, the RD obtained some valuable information though much of the time was wasted. The obvious conclusion was the dominance of SP's mother.
****

QUESTIONS CONTINUED:

11. How would you handle SP's mother in the above situation?

12.   If blood was drawn and analyzed, what results would you expect to find? Explain your expectations.

Refeeding someone after a period of starvation should be done with caution. A starved person does not just start eating large meals. Generally, the longer the starvation period, the slower the refeeding. Each case is different and should be evaluated on its own circumstances. SP has been dieting for about a year, the last several months being a starvation diet. She lost from 120 to 85 lbs.

The psychologist, RD, and physician worked as a team with SP's father and convinced him that SP needed to have more of a voice in her life. In turn, he convinced his wife. They agreed to allow SP more freedom. They gave her their blessing to change her major to special education. This greatly changed the home atmosphere and gave the psychologist and RD a chance to work with SP. By now she had created a tremendous fear of being fat and did not want to gain much weight. They were able to show her that her dieting hurt her social life more than it hurt her parents. They also helped her to dream of being the teacher and role model she wanted to be for children. She slowly started to eat more but gained several pounds very fast. She was afraid she would continue to gain weight at that rate and would get fat.

The RD explained the occurrence of rapid weight gain to her and calmed her fears. SP continued to gain weight. As she did, she began to feel stronger. Still fearing to gain too much, she restarted her vigorous exercise routine. She remembered how much she enjoyed eating and began to overeat, using the excessive exercise as an excuse. She enjoyed what she was doing but felt guilty for overeating. She gained 30 lbs back and was now 115 lbs. SP was released from the care of her physician and counselors.

She started to binge and purge often. She was in a real dilemma. She wanted to eat but did not want to gain weight. She was willing to purge to keep her weight down but she found that gross. She did not like sticking her finger down her throat. One day a friend introduced her to ipecac syrup, an over-the-counter drug that would make her vomit. SP began using this drug with laxatives and diuretics to keep her weight down while overeating.
****

QUESTIONS CONTINUED:

13.   Assume that SP will submit to ending her starvation diet. Calculate SP's BMR using the Harris-Benedict equation. What activity factor would you initially use? Outline a refeeding plan for SP and include in your plan: the number of kcals you would start with per day, the rate at which you would advance, and your final kcals per day.

14. If you feed too much too fast, what problems would you expect SP to have?

bulky foods can cause gastrointestinal intolerance

15. Explain why there would be a rapid weight gain right after going off a starvation diet.

16. How would you counsel SP behaviorally?

help her become an effective independently person. Convey principles rather than rigid rules. Positive, regular habits should be encourgage.

17. List the characteristics of bulimia.

① reacurring episodes of binge eating
② sense of a lack of control
③ self evaluation unduly influenced by weight & body shape
④ vomming, use of laxatives fasting, excessive exercise

18. List the complications of bulimia.

19. Describe the action, nutritional complications, and any adverse reactions of ipecac syrup.

20. What are the functions of laxatives and diuretics?

21. If SP took ipecac syrup, laxatives, and diuretics on a daily basis, predict the probable complications. What lab values would most likely be affected? Be specific and explain.

laxatives and diuretic abuse can cause cardiac arrest and other problems. Ipecac syrup Nausea, vommiting and constipation can occur

22. Discuss the role of ipecac syrup, laxatives, and diuretics in causing weight loss.

****

SP thought she could now binge and cause herself to lose the food with her over-the-counter drugs and no one would know. Her binging increased. On some days she might throw up once or twice or ten times. A typical binge could include one large bag of chocolate chip cookies, one liter of soda, two peanut butter sandwiches, a large bag of potato chips, almost a half-gallon of ice cream, and several candy bars. She might throw up four or five times during this binge. Sometimes she would spend $30.00 to $40.00 a day for binge food.

SP went to her dentist for a semi-annual check up. Upon examining her mouth, he noted perimolysis. When he asked her if she had been vomiting a lot, she denied it.
****

QUESTIONS CONTINUED:

23. Describe the approach you would use to counsel SP now that she is bulimic. Would your nutritional goals change? Explain.

24. What is perimolysis? Explain its relationship with bulimia.

25.   Summarize anorexia and bulimia by listing all the possible characteristics of both in column
      one below; in column two, place a check by those that pertain to anorectics; in column
      three, place a check by those that pertain to bulimics. Note those characteristics that are
      common to both by circling them.

CHARACTERISTICS	ANOREXIA	BULIMIA
Starvation	✓	✓
Binge & purge		✓
exercise	✓	✓
Obsessive	✓	✓

26.   Using any of the given information, compose a SOAP note about SP.

S: _____

_____

_____

_____

O: _____

_____

_____

_____

A: _____

_____

_____

_____

P: _____

_____

_____

_____

Related Readings

1. Davis, C., Kaptein, S., Kaplan, A.S., Olmsted, M.P., & Woodside, D.B. (1998). Obsessionality in anorexia nervosa: the moderating influence of exercise. *Psychosom. Med.* 60(2):192-197.

2. Position paper of the American Dietetic Association. (1994). Nutrition intervention in the treatment of anorexia nervosa, bulimia nervosa, and binge eating. *J. Am. Diet. Assoc.* 94:902.

3. Position paper of the American Dietetic Association. (1997). Weight management. *J. Am. Diet. Assoc.* 97:71.

4. Pryor, T., & Wiederman, M.W. (1998). Personality features and expressed concerns of adolescents with eating disorders. *Adolescence.* 33(130):291-300.

5. Siegel, M., Brisman, J. & Weinshel, M. (1989). *Surviving an Eating Disorder.* New York: Harper & Row.

6. Sullivan, P.F., Bulik, C.M., Carter, F.A., Gendall, K.A., & Joyce, P.R. (1996). The significance of a prior history of anorexia in bulimia nervosa. *Int. J. Eat. Disord.* 20(3):253-261.

7. Whitney, E.N., Cataldo, C.B., & Rolfes, S.R. (1998). *Understanding Normal and Clinical Nutrition*, 5th Ed. West/Wadsworth.

8. Wills-Brandon, C. (1989). *Eat Like a Lady. Guide for Overcoming Bulimia.* Deerfield Beach, Flordia. Health Communications, Inc.

CASE STUDY #6
POLYPHARMACY

INTRODUCTION
This case study describes a typical situation that occurs often, particularly with the elderly. It involves several complications that by themselves are not serious, but together lead to disaster. Several medications are included in this study and the Nutritional Screening Initiative is introduced.

SKILLS NEEDED

ABBREVIATIONS:
Knowledge of the following abbreviations is required in order to understand this case. You should learn these abbreviations before you begin to read the study.

ALT : alanine aminotransferase
AST : aspartate aminotransferase
b.i.d : twice a day
BMI : body mass index
BUN : blood urea nitrogen
Chol : cholesterol
Cl : chloride
C/O : complains of
Cr : creatinine
D/C : discontinue
Dx : diagnosis
g : gram
g/dl : grams per deciliter
Glu : glucose
Hct : hematocrit
Hgb : hemoglobin
hs : hour of sleep or evening
Ht : height
I.V. : intravenous
K : potassium
MCV : mean corpuscular volume
mEq/L : milliequivalent per liter

mg : milligram
mg/dl : milligram per deciliter
mm^3 : cubic millimeter
mos : months
Na : sodium
po : by mouth
prn : as needed
PVC : premature ventricular contraction
qod : every other day
q6h : every six hours
RBW : reference body weight
Ser Alb : serum albumin
t.i.d. : three times a day
UBW : usual body weight
U/L : units per liter
UTI : urinary tract infection
WBC : white blood cell count
Wt : weight
YOWF : year old white female
μm^3 : cubic micromoles
¼NS : one fourth strength normal saline

FORMULAS:
The formulas used in this case study are: reference body weight and percent reference body weight (Appendix A, Tables A - 7 through A - 10).

LABORATORY VALUES:
You will need to be able to interpret the nutritional significance of the following laboratory values for

this case study: BUN, Cl, Cr, Glucose, Hct, Hgb, K, Lymphocytes, MCV, Na, Ser Alb, ALT, AST, Chol, and WBC (Appendix B).

MEDICATIONS:

Become familiar with the following medications before reading this case study. Note the diet-drug interactions, dosages and method of administration, gastrointestinal tract reactions, etc.

1. Feldene (piroxicam); 2. FML Liquifilom Ophthalmic; 3. Maxitrol Ointment; 4. Aldomet (methyldopa); 5. Lasix (furosemide); 6. Norpace (disopyramide phosphate); 7. Dalmane (flurazepam hydrochloride); 8. Ativan (lorazepam); 9. Gantanol (sulfamethoxazole); 10. Tagamet (cimetidine); 11. Aquaphyllin (theophylline).

Mrs. SJ is a 75 YOWF who has been a widow for 15 years. Until recently, she had been active in various groups such as the "Quilters" and the "Crafters." Arthritis in her hands and arms started making it increasingly difficult for her to sew or make crafts. Arthritis in her knees and hips also made it difficult for her to move freely. She has to walk with a cane. Her Family Practice physician sent her to a rheumatologist. She has been taking 20 mg of Feldene (piroxicam) po once daily for a number of years.

Mrs. SJ smoked a pack of cigarettes a day for 40 years before being forced to quit by her physician and family. A severe cough, difficulty breathing, and the diagnosis of chronic pulmonary emphysema also helped to convince her to quit. She has been taking theophylline for three years for this condition.

Mrs. SJ has also been having problems with her eyes. Not only does she have to get her glasses changed more frequently, but her eyes have been very watery and burn a lot. Her doctor has not determined if she has an allergy or if she has an infection. She has not yet responded to treatment. Her ophthalmologist prescribed FML Liquifilm Ophthalmic for allergies and Maxitrol Ointment for infection. Ophthalmology visits are expensive and are not covered by her health insurance.

Mrs. SJ has been underweight all her life even though she had a ravenous appetite until recently. She is a 5'5" small-framed woman and weighs 100 lbs. In her prime, the most she ever weighed was 111 lbs with very little variation from year to year. This is the least she has weighed since high school. Since about age 60, she has been having a problem with hypertension and has been taking Aldomet (methyldopa), 500 mg b.i.d., po. She had a small but noticeable amount of edema in her feet and C/O her rings being too tight. Her doctor prescribed Lasix (furosemide) 20 mg qod prn. Mrs. SJ started having arrhythmias with frequent PVCs around age 65 and had to see a cardiologist. He prescribed Norpace (disopyramide phosphate), 150 mg, q6h, po.

Mrs. SJ became depressed about a year ago. Several factors brought this about. The death of her older sister was a shock to her. Mrs. SJ was the fourth of five children but now is the oldest alive. She still has a younger brother who lives 200 miles away. She has two children but they both moved to California and Mrs. SJ lives on the East coast. After her sister died, two of her closest friends in her craft club passed away within three months of each other. They were both younger than she is and used to be her means of transportation since they both had cars and could drive well. She fears that she is now the next to die and that thought does not please her.

All the deaths took place during the winter when Mrs. SJ has the most trouble with her arthritis.

The past winter was a particularly bad one with a lot of snow on the ground which meant Mrs. SJ was stuck indoors for weeks at a time. She lives in an efficiency apartment in a complex inhabited by elderly tenants.

The depression has been due to the added stress and this has aggravated her arrhythmias. Her heart rate is usually good but, when the arrhythmias occur, Mrs. SJ becomes very weak and dizzy. At these times she does not want to go anywhere for fear of having an attack. The more fearful she becomes, the more the arrhythmias occur. This makes her even more fearful. It also increases her blood pressure, which adds to the stress. Another problem of Mrs. SJ's was gastritis. After years of complaining about a burning sensation in her stomach, she consulted a gastroenterologists. A mild case of gastritis was diagnosed. Most of the time this does not bother her but, with the added stress, the burning sensation returned. The gastroenterologists prescribed cimetidine which Mrs. SJ has been taking for years.

This downward spiral has been going on for several months and affects her sleep. She did not tell her Family Practice physician everything, but she told her about not sleeping. Since her rest is important, her Family Practice physician gave her 15 mg of Dalmane (flurazepam hydrochloride) at hs for sleep and Ativan (lorazepam), 1 mg t.i.d. for anxiety. Neither of these seemed to work fast enough for her so she doubled the doses. The additional stress also caused her to have anorexia which resulted in weight loss and weakness. All of this lowered her immune system and she ended up with a bladder infection. For this her physician gave her Gantanol (sulfamethoxazole), 1 g po, .b.i.d. and told her to force fluids.

Mrs. SJ felt too weak and sick to go to the corner grocery to buy food and too sick to prepare it. She stayed inside and ate cereal and milk until she ran out of milk. She ate soft foods like rice with a little margarine and hot tea. After several days of this, she became too weak to get out of bed. When she did not show up in the recreation room for Friday night BINGO, something she never missed, some of the tenants went to check on her. She was found in bed very confused and disoriented and could not even sit up. They called 911 and had her sent to the hospital. Mrs. SJ was admitted with a Dx of UTI, dehydration, anemia, and depression. Her admission lab values were as follows:

TEST	RESULT	NORM	TEST	RESULT	NORM	TEST	RESULT	NORM
Hgb	17.0 g/dl	12 - 16	MCV	78μm^3	82 - 98	Chol	130 mg/dl	140 - 190
Hct	50%	36 - 48	WBC	14 0 x 10^3/mm^3	5 - 10 x 10^3/mm^3	Na	148 mEq/L	135 - 145
K	5.2 mEq/L	3.5 - 5.3	Cl	106 mEq/L	98 - 106	AST	85 U/L	5 - 40
BUN	15mg/dl	7 - 18	Cr	0.8mg/dl	0.6 - 1.3	Ser Alb	4.3 g/dl	3.9 - 5.0
Lympho cytes	8%	20 - 40	Glu	160 mg/dl	65 - 110	ALT	20 U/L	7 - 56

****

QUESTIONS:

1. Define:
    Rheumatologist:

    Arthritis:

    Ophthalmologist:

    Cardiologist:

    Arrhythmias:

    Anorexia: *characterized by refusal to maintain minimum*

    Force Fluids: *incuraging the patient to drink alot*

2. Mrs. SJ is taking a lot of medications. This is typical for the elderly and is cause for concern, particularly when they are seeing more than one physician and not telling each physician what the others are prescribing. Research each medication Mrs. SJ is taking and list the possible nutritional complications for each.

MEDICATION	ACTION	NUTRITIONAL COMPLICATION
Feldene (piroxicam)	*stomach ulsers, fluid retention*	
FML Liquifilom Ophthalmic		
Maxitrol Ointment		
Aldomet (methyldopa)	*Dry mouth, diarrhea, sedation, drowsiness, headache peripheral edema, fever, lowers blood pressure*	
Lasix (furosemide)	*Oral irritation, stomach cramps, possible hypotention dizziness, blurred vision*	
Norpace (disopyramide phosphate)	*headache, weakness, photosensativity dry mouth, diarrhea, blurred vision, headache*	
Dalmane (flurazepam hydrochloride)	*dizziness, weakness, fatigue*	
Ativan (lorazepam)	*Dry mouth, nausea, constipation, sedation, dizziness, weakness, ataxia*	
Gantanol (sulfamethoxazole)	*confusion, depression, memory impairment*	

3. Each of these medications is given to correct a problem but each could possibly cause a
    side effect that may negate the effects of another medication or increase the intensity of
    another symptom. Example: Piroxicam is a nonsteroidal anti-inflammatory for mild to
    moderate pain caused by osteoarthritis or rheumatoid arthritis. Possible side effects include
    peptic ulcer, anxiety, palpitations, dysrhythmias, and insomnia. List the possible side effects
    that could occur with the following medications.

MEDICATION	INTERACTION WITH OTHER MEDS
Feldene (piroxicam)	
FML Liquifilom Ophthalmic	
Maxitrol Ointment	
Aldomet (methyldopa)	
Lasix (furosemide)	
Norpace (disopyramide phosphate)	
Dalmane (flurazepam hydrochloride)	
Ativan (lorazepam)	
Gantanol (sulfamethoxazole)	

4. Discuss Mrs. SJ's symptoms in light of her medications. Do you think some of her
    problems could have been caused by drug interactions? Explain. *I would think the
    weakness from the medications certainly had an effect on
    her imobility. She is definately dehydrated from the
    medications because of the diarrhea, and dry mouth.
    Even some of the medications can cause depression, although
    her being a widow for 15 years may also contribute.*

5. Considering her diseases and medications, describe an adequate home diet for Mrs. SJ.
    Interject in your discussion nutrients that she needs to include and substances she should
    avoid or ingest sparingly.

****

When Mrs. SJ's physician saw all of the drugs she was taking from four different doctors, she told her that one of her problems was "polypharmacy." She D/C'd all of her at-home medications and started I.V. antibiotics with ¼NS and an antihypertensive medication. All other medications were given on a prn basis. Within a couple of days, Mrs. SJ's infection was under control and she was awake and alert. The dietitian visited her to do an assessment using Level I of the Nutrition Screening Initiative and obtained the following information:

♦ **BODY WEIGHT HISTORY**

▸ **Ht 5'5"   Wt 100 lbs**

▸ **Wt 1 yr ago 111 lbs**

▸ **Wt 6 mos ago 111 lbs**

The following statements on the Level I Screening Initiative were all answered positively by Mrs. SJ.

♦ **EATING HABITS**

- ▸ **Usually eats alone**
- ▸ **Has poor appetite**
- ▸ **Eats vegetables two or fewer times daily**
- ▸ **Eats milk products once or not at all daily**
- ▸ **Eats fruit or drinks fruit juice once or not at all daily**
- ▸ **Eats breads, cereals, pasta, rice, or other grains five or fewer times daily**
- ▸ **Has difficulty chewing or swallowing**

♦ **LIVING ENVIRONMENT**

- ▸ **Lives alone**
- ▸ **Is housebound**

♦ **FUNCTIONAL STATUS**
Usually or always needs assistance with

- ▸ **Walking or moving about**
- ▸ **Traveling (outside the home)**
- ▸ **Preparing food**
- ▸ **Shopping for food or other necessities**

****

6. Obtain a guideline for the Nutrition Screening Initiative (available from: Nutrition Screening Initiative, 2626 Pennsylvania Avenue, N.W., Suite 301, Washington, D.C. 20037 or contact your Ross Laboratories representative). Using this guide, identify the agencies that should be contacted for each of the problems in the outline above. With the Nutrition Screening Initiative as a guide, discuss possible solutions to the problems mentioned above.

7. Consider other problems Mrs. SJ has, such as money for medications and physicians and all the causes for depression, etc. Discuss how these problems have to be addressed if Mrs. SJ is to improve. List the possible solutions.

8. What did the physician mean by "polypharmacy"?

Several catagories of medications can alter the integrity of oral mucus, taste sensation, and salivory production. Our responsibility to assess problems with biting, chewing or swallowing or presenses of sores in the mouth

9. What is Mrs. SJ's RBW and percent of RBW (Appendix A, Tables A - 7 through 10)? Please show all work.

132

$$\frac{100}{132} \times 100 = 75\%$$

10. What is her UBW and percent UBW (Appendix A, Table 10)?

111

$$\frac{100}{111} \times 100 = 90\%$$

11.   What is Mrs. SJ's BMI (Appendix A, Tables A, 12 and 13)? Would her age affect her BMI?

$$\frac{100}{(5.5)^2} = \frac{100}{30.25}   3.3\ BMI$$

****

After a few more days, Mrs. SJ's labs were repeated and the following obtained:

TEST	RESULT	NORM	TEST	RESULT	NORM	TEST	RESULT	NORM
Hgb	11.0 g/dl	12 - 16	MCV	$80\mu m^3$	82 - 98	Chol	140 mg/dl	140 - 190
Hct	35%	37 - 48	WBC	$7\,0\ x\ 10^3/mm^3$	$5 - 10\ x\ 10^3/mm^3$	Na	139 mEq/L	135 - 145
K	3.8 mEq/L	3.5 - 5.3	Cl	104 mEq/L	98 - 106	AST	40 U/L	5 - 40
BUN	7mg/dl	7 - 18	Cr	0.6mg/dl	0.6 - 1.3	Ser Alb	3.5 g/dl	3.9 - 5.0
Lympho cytes	19%	20 - 40	Glu	90 mg/dl	65 - 110	ALT	55 U/L	7 - 56

****

Questions 12 - 16 require a greater knowledge of the interpretation of lab values.

12.   Using the first set of lab values, list the parameters that are out of the normal range and give the possible explanations for this.

13.   Do the same for the second set of lab values.

14.   Compare the sets of lab values and explain why the change occurred parameter for which there was a significant change, namely: Hgb, Hct, K, BUN, Lymphocytes, WBC, glucose, Na, AST, serum albumin, and ALT.

15.   What could be concluded from the low Cr level in the second set of lab values? (☞ *Hint: Consider Mrs. SJ's weight.*)

The low cr level can also show that she is not getting enough meat in her diet, as says she even has difficulty in swallowing. Her energy storage is depleated, and her muscles are becoming alot weeker.

16.   Explain how Mrs. SJ's lifestyle and diet could have contributed to some of the results in the first set of lab values.

17.   Considering Mrs. SJ's anthropometrics, medical history, and nutritional history, outline a patient care plan for her. Be sure to include a teaching plan.

18.   If you were the dietitian assessing Mrs. SJ, write an appropriate SOAP note based on the information stated above.

S: 75 year old white female. Arthritis dificult to move, lost ability to sew. Depression. Difficulty in her intake of food due to the heavy medications

O: underweight, lost 11 pounds within a year. A bad appetite, She eats alone. Does not like to drink juice, and does not like milk

A: Very poor condition. Patient is under IBW, medications all have the same side effects, should be taken off all medication

P:

Related Reading

1. Atkin, P.A., Stringer, R.S., Duffy, J.B., Elion, C., Ferraris, C.S. Misrachi, S.R., & Shenfield, G.M. (1998). The influence of information provided by patients on the accuracy of medication records. *Med. J. Aust.* 169(2):85-88.

2. Beers, MH, Munekata, M, & Storrie, M. (1990). The accuracy of medication histories in the hospital medical records of elderly persons. *J. Am. Geriatr. Soc.* 38(11):1183-1187.

3. Chernoff, R. (1991). *Geriatric Nutrition.* 3rd Ed. Gaithersburg, Maryland. Aspen Publishers

4. Fischbach, F.T. (1995). *A Manual of Laboratory & Diagnostic Tests.* 5th Ed. Philadelphia. J.B. Lippincott Company.

5. Grant, A. & DeHoog, S. (1991). *Nutritional Assessment and Support.* 4th Ed. Seattle, Washington. A. Grant and S. DeHoog, Publishers.

6. Lamy, P.P. (1987). Age-associated pharmacodynamic changes. *Methods Find Exp Clin Pharmacol.* 9(3):153-159.

7. Lee, R.D. & Nieman, D.C. (1993). Nutritional Assessment. Brown and Benchmark, pp. 180-188.

8. Nutrition Screening Initiative. (1991). A project of the American Academy of Family Physicians, the American Dietetic Association, and the National Council on Aging, funded in part by Ross Laboratories. 2626 Pennsylvania Ave., NW, Suite 301, Washington, DC 20037.

9. Position paper of the American Dietetic Association. (1996): Nutrition, aging, and the continuum of care. *J. Am. Diet. Assoc.* 96:1048.

10. Position paper of the American Dietetic Association. (1998): Liberalized diets for older adults in long-term care. *J. Am. Diet. Assoc.* 98:201.

11. Prendergast, A. & Fulton, F.L. (1997). Medical Terminology: A Text/Workbook. 4th Ed. Redwood City, California. Addison-Wesley Nursing.

12. Price, D., Cooke, J., Singleton, S., & Feely, M. (1986). Doctors' unawareness of the drugs their patients are taking: a major cause of overprescribing? *Br. Med. J.* 292(6513):99-100.

13. Pronsky, Z.M. & Solomon, E. (1997). *Food-Medication Interactions.* 10th Ed. Phoenix, Arizona. Food-Medication Interactions, Publishers and Distributors.

14. Schlenker, E.D. (1997). *Nutrition in Aging.* St. Louis, Missouri. Mosby - Year Book, Inc.

15. Torrible, S.J., & Hogan, D.B. (1997). Medication use and rural seniors. Who really knows what they are taking? *Can. Fam. Phy.* 43:893-898.

16.   Troncale, J.A. (1996). The aging process. Physiologic changes and pharmacologic implications. *Postgrad Med.* 99(5):111-114.

17.   White, J.V.,Ham, R.J., Lipschitz, D.A.,Dwyer, J.T., & Wellman, N.S. (1991). Consensus of the Nutrition Screening Initiative: risk factors and indicators of poor nutritional status in older Americans. *J. Am. Diet. Assoc.* 91:783-787.

18.   White, J.V.,Ham, R.J., & Lipschitz, D.A. (1991). *Report of Nutrition Screening 1: Toward a Common Goal.* Washington, DC: Nutrition Screening Initiative.

19.   White, J.V., Dwyer, J.T., Possner, B.M., Ham, R.J. Lipschitz, D.A., & Wellman, N.S. (1992). Nutrition Screening Initiative: development and implementation of the public awareness checklist and screening tools. *J. Am. Diet. Assoc.* 92:163-167.

20.   Whitney, E.N., Cataldo, C.B., & Rolfes, S.R. (1998). *Understanding Normal and Clinical Nutrition*, 5th Ed. West/Wadsworth.

CASE STUDY #7
OBESITY TREATED WITH DIET

INTRODUCTION
This is the first of several studies that specifically pertain to the treatment of obesity in adults. Each study involves a different mode of treatment. In this study, dietary modification and exercise, the preferred components of treatment, are employed.

SKILLS NEEDED

ABBREVIATIONS:
Knowledge of the following abbreviations is required in order to understand this case. You should learn these abbreviations before you begin to read the study.

BMI : Body mass index
Chol : cholesterol
lbs : pounds
Na : sodium
RBW : Reference body weight

RD : registered dietitian
Sat Fat : saturated fat
Tot Fat : total fat
YOWM : year old white male

FORMULAS:
The formulas used in this case study include reference body weight and percent reference body weight (Appendix A, Tables A - 7 through A - 10). Total energy needs, protein needs, and activity factors can be found in Appendix D, Tables D - 1 and D - 2.

RB is a 45 YOWM who is married with two children, ages 15 and 18. He is of Italian descent and loves to eat Italian food, although any kind of food will do. He takes pride in his ability to prepare Italian food. RB is a high school mathematics teacher who does not exercise. His love of food and lack of exercise has created a problem: he is overweight. For years he has endured much criticism from his family and friends about his health. His family history reveals several areas of concern. His father died of a heart attack and two brothers have had heart attacks. He has another brother and a sister who have high serum cholesterol and are on strict diets. Both of his parents and all of his grandparents had problems with high serum cholesterol. None of this really bothered him until recently. Since his last birthday he has become very concerned about his health and has decided to do something about his weight.

He is 5'10" tall and weighs 215 lbs. Throughout college he weighed 170 lbs and, according to his wife, "looked good". She wants him to look like that again, so he has decided to lose 45 lbs and get back to his college weight. He has even been talking about starting an exercise program and getting back into his college running condition. Because of his age and his strong family history of cardiovascular disease, his wife convinced him to obtain medical advice first. The family physician found him to be healthy except for his weight. The physician reinforced what the others had been telling him about losing weight and exercising more. He emphasized that he should start slowly and sent him to the clinic RD to discuss his weight problem.

The RD interviewed him and obtained the following information:

RB sleeps as late as he can and is usually in a rush in the morning. He occasionally eats breakfast. He drinks a cup of coffee (with sugar) while getting ready for work. He does not eat in the school cafeteria at lunch because he does not like the food. He says it is not real food. The RD discovered that RB does not eat for the sake of eating or boredom; he eats because he enjoys good food. He brings a sandwich from home and drinks a soda. His sandwiches are usually egg salad, chicken salad, deli roast beef, or salami, made with rye, whole wheat, or sometimes French bread. He puts some mustard and a little bit of mayonnaise with some lettuce and tomato on his sandwiches.

He does not eat again until dinner. He likes to go home and work on his class work before eating. When he sits at the table in the evening, he wants to have the day behind him and nothing left to do but enjoy a good meal and go to bed. He usually helps prepare dinner, or he prepares it himself. This is the meal he lives for. The way RB sees it, he works hard during the day, skips breakfast, eats a light lunch, and does not snack much during the day, so he deserves a big meal at night.

RB has never been to Italy but has researched the lifestyle and eating habits there. He has selected those recipes he likes and has combined them with some traditions of this country to come up with his own style of Italian food. The meal has to start with the American traditional tossed salad with about 1/4 c of olive oil and vinegar dressing. The Italian tradition of pasta appears at each evening meal. Spaghetti, mostaccioli, fettuccini, or another form of pasta is always included. Tomato sauce, grated cheese, or cheese sauces are essential. Parmesan, Romano, and mozzarella are his favorites. RB and his wife use some chicken and shellfish, but beef and veal are more popular. Lots of Italian seasoning, salt, and olive oil are important. Garlic bread is included in each meal with olive oil and a touch of melted butter. RB insists on real butter for his cooking. He also likes to use real cream in his sauces. Egg batter helps to keep things together. The vegetables are usually eggplant, squash, artichokes, asparagus, or beets. He uses a lot of onions, garlic, bell pepper, celery, and parsley in his cooking. All of this is washed down with several glasses of wine and finished off with pistachio ice cream or an Italian pastry. The serving sizes vary a great deal, depending on the combination of foods served at one meal. However, excessive amounts of each food item are the norm rather than the exception.

After eating there is some cleaning up to be done before bedtime. Weekend meals are not that much different except that the Sunday meal is usually earlier.

****

QUESTIONS:

1. Determine RB's RBW and percent RBW (Appendix A, Tables A - 7 through 10). Please show all calculations for this report.

2. Using the Harris-Benedict equation and the appropriate activity factor, calculate his daily energy needs (Appendix D, Tables D - 1 and D - 2).

3. Determine his protein needs (Appendix D, Table D - 2).

4. Determine his BMI and interpret the results in relation to health risks (Appendix A, Tables A - 12 and 13).

5. If RB were to lose one pound per week, he would have to reduce his daily caloric intake by how many kcals?

6. Explain the importance of exercise in RB's weight reduction plan. Include advice you would give RB about the elements of an exercise program, including starting a program, warming-up, duration, intensity, etc.

7.   Give some examples of behavior changes you would recommend to RB to promote a healthier lifestyle. This would include nutrition, exercise, daily schedule, etc.

8.   Briefly discuss the importance of RB's family medical history.

9.   The RD's observation of the reason RB eats is important. Discuss how you would use this information in counseling RB.

10.   In column one, list each food in RB's diet that is high in a potentially harmful nutrient. In columns 2 - 6, put a check by the foods that are high in these constituents. In column 7, give an example of a potentially healthier food that can be substituted for the food in column 1.

FOOD / TOT FAT / SAT FAT / CHOL/ Na / SUGAR/   SUBSTITUTE

11. Using the information obtained above, evaluate RB's diet for nutrient deficiencies and/or excesses.

12. RB believes it is all right to skip breakfast, eat light during the day, and eat a big meal at night. Is anything wrong with this kind of thinking? If so, identify the problem and describe the solution.

13.   Plan a day's menu for RB, including kinds of food, amounts, and times eaten.

Related Readings

1.   Berg, F.M. (1993). Health Risks of Obesity. 2nd Ed. Obesity & Health, Healthy Living Institute, Hettinger, N.D.

2.   Chisholm, D.J., Samaras, K., Markovic, T., Carey, D., Lapsys, N., & Campbell, L.V. (1998). Obesity: genes, glands or gluttony? *Reprod. Fertil. Dev.* 10(1):49-53.

3.   Green, S.M. (1997). Obesity: prevalence, causes, health risks and treatment. *Br. J. Nurs.* 6(20):1181-1185.

4.   Mayer, J. (1966). Some aspects of the problem of regulation of food intake and obesity. *N. Engl. J. Med.* 274(11): 610-616.

5.   NHBLI (1998) *Clinical Guidelines on the Identification, Evaluation, and Treatment of Overweight and Obesity in Adults.* NHLBI Information Center, P.O. Box 30105, Bethesda, MD 20824-0105.

6.   Position paper of the American Dietetic Association. (1997). Weight management. *J. Am. Diet. Assoc.* 97:71.

7.   Weight-control Information Network. November, 1993. *Understanding Adult Obesity.* NIH Publication No. 94-3680.

8.   Weight-control Information Network. March, 1995. *Weight Cycling.* NIH Publication No. 95-3901.

9.   Weight-control Information Network. May, 1998. *Do you know the health risks of being overweight?* NIH Publication No. 98-4098.

10.   Weinsier, R.L., Hunter, G.R., Heini, A.F., Goran, M.I., & Sell, S.M. (1998). The etiology of obesity: relative contribution of metabolic factors, diet, and physical activity. *Am. J. Med.* 105(2):145-150.

11.   Whitney, E.N., Cataldo, C.B., & Rolfes, S.R. (1998). *Understanding Normal and Clinical Nutrition,* 5th Ed. West/Wadsworth.

CASE STUDY #8
OBESITY TREATED WITH SURGERY

INTRODUCTION
This study, like Case Study #7, is also concerned with obesity, but the treatment in this case is centered around surgery rather than diet. Surgery is intended for those who are morbidly obese and cannot obtain weight reduction with diet alone. Most dietitians do not favor this type of treatment for obesity. The old adage, "an ounce of prevention is worth a pound of cure" is the preferred approach. Another way of saying this is to have a "healthy lifestyle" that includes proper diet and exercise to prevent obesity. However, once someone becomes morbidly obese, it is very difficult to lose weight. Exercise may not be a possibility. If the obesity becomes life threatening, some resort to surgery. It is doubtful that surgery will help, but as long as it is available, there are some who will chose this option. Agreeing or disagreeing with the procedure is not the main point of this case study. If there are those who will have this surgery, then dietitians need to understand the risks, complications, and treatment so that they can provide the patient with the "best" nutritional follow-up.

SKILLS NEEDED

ABBREVIATIONS:
Knowledge of the following abbreviations is required in order to understand this case. You should learn these abbreviations before you begin to read the study.

BMI ; body mass index
BUN : blood urea nitrogen
c : cup
Ca : calcium
Cl : chloride
Cr : creatinine
FH : family history
g/dl : grams per deciliter
Hct : hematocrit
Hgb : hemoglobin
K : potassium
Lymph : lymphocytes
MCV : mean corpuscular volume
mEq/L : milliequivalent per liter

mg/dl : milligram per deciliter
MH : medical history
mm^3 : cubic millimeter
Na : sodium
oz : ounce
RBW : reference body weight
RD : registered dietitian
Ser Alb : serum albumin
SH : social history
SOB : shortness of breath
T : tablespoon
tsp : teaspoon
WBC : white blood cell count
YOWM : year old white male
μm^3 : cubic microns

LABORATORY VALUES:
You will need to be able to interpret the nutritional significance of the following laboratory values for this case study: BUN, Cl, Cr, glucose, Hct, Hgb, K, lymphocytes, MCV, Na, and ser alb (Appendix B).

FORMULAS:
The formulas used in this case study include reference body weight, adjusted body weight (Appendix A, Tables A - 7 through 11), the Harris-Benedict equation, activity factors, and protein requirements (Appendix D, Tables D - 1 and D -3).

Mr. Y is a 39 YOWM with morbid obesity. He is 6'1" and weighs 320 lbs. He has been overweight all his life, even as a child. He and his doctor have noticed that there is a definite correlation between his increase in age and his increase in weight. He has tried numerous diets without success. He has lost a significant amount of weight with some diets, but they were "so monotonous" he could not stay on them. When he went off a diet, it seemed like he would eat that much more, like he was trying to catch up on the food he missed. He is an accountant with a desk job, one at which he can eat all day long without opportunity for exercise. He has never had an exercise program.

His doctor has warned him that if he does not do something about his weight, he will not live to see his grandchildren. Mr. Y has been noticing an increased SOB and difficulty accomplishing simple physical tasks. Even though his FH is good concerning heart attacks, strokes, diabetes, cancer, etc., his doctor informed him that he is at a much higher risk for these diseases. He also pointed out that no one in his family is as big as he is. Therefore, Mr. Y has decided to take drastic measures to lose weight. He has agreed to have gastric partitioning surgery. His doctor has explained the risks and benefits of the surgery. He referred him to a surgeon who has a good record with bariatric surgery. He was admitted to the hospital on Monday and was to have surgery on Wednesday. The physician asked the RD to visit him and begin teaching him his new diet. The RD read his chart and made note of his FH, SH, and MH.

His lab values were not significant:

TEST	RESULT	NORM	TEST	RESULT	NORM	TEST	RESULT	NORM
Hgb	17.0 g/dl	14-17	BUN	20 mg/dl	7 - 18	MCV	89 μm³	82 - 98
Hct	48%	42 - 52	Cr	1.3 mg/dl	0.6 - 1.3	Lymph	17.0%	20 - 40
Ser Alb	4.4 g/dl	3.9 - 5.0	Na	133 mEq/L	135 - 145	K	4.2 mEq/L	3.5 -5.3
Glucose	124 mg/dl	65 - 110	Cl	95 mEq/L	98- 106	WBC	10.2x10³ /mm³	5 - 10 x 10³

The RD went into Mr. Y's room to interview him and to determine his usual intake. She obtained the following information:

Mr. Y claimed that he actually does not eat that much. He gets up at 7 A.M. and eats breakfast with his wife. Usually this consists of:

- 2 fried eggs
- 1 glass of orange juice (8 oz)
- 2 pieces of toast
- 1 T jelly per slice
- 1 c grits
- 1 pat of margarine per slice
- 2 c of coffee with 1 tsp of cream and 2 tsp of sugar per cup

Once at work, he snacks on food all day. He always has coffee (the usual cream and sugar) or iced tea (with 2 tsp of sugar per 8 ozs) on his desk and "sips" all day. In the morning, glazed doughnuts are usually available. Cheese crackers are one of his favorite snacks, and every now and then, a candy bar appeals to him. Sometimes he brings a big can of roasted peanuts or a bag of candy, like chocolate-covered raisins, and eats "one or two every now and then." By lunch he has worked up an appetite. He and his co-workers usually eat at the sandwich shop across the street. He likes the sloppy roast beef or club special. He must have french fries with a sandwich, the two "sort of just go together." He said he knows that you need fiber in your diet, so a salad is the norm also, with about 1/4 c of french dressing. He may have a dessert that consist of home-made pie and/or a scoop of ice cream for calcium.

He follows the same routine in the afternoon as he does in the morning. At suppertime, his wife has a big meal for him. This consists of salad for fiber with 1/4 c of salad dressing. He has at least 4 to 5 ounces of meat, as roast with gravy, fried chicken, or fried fish if it's Friday. Sometimes beef steak, ham steak, or fried pork chops are served. Potatoes, rice, or macaroni are always included as the starch ("about 1 cup or so"). She always has a vegetable. Carrots, green beans, and squash are typical. These are usually prepared with salt and butter, fat back, or bacon. Several slices of bread and butter are always served. The drink is tea, sweetened with 2 tsp of sugar per 8 ozs. Several glasses per meal is not uncommon. He is still not "big on desserts" but, when his wife fixes something, he eats his share. Mr. Y eats his food very fast, taking little time to chew properly. He enjoys his food so much that he wants to eat it as fast as possible. This is not only true for his evening meal but for everything he eats. This is a habit he developed as a child.

After supper he is really not that full, so as he sits in front of the TV and relaxes with the newspaper he continues to munch on popcorn (salted, with lots of melted butter), peanuts, cookies, or whatever is in the house. A couple of beers always helps him relax, but he only drinks light beer because it is "less filling." It does not take too long for Mr. Y to become very sleepy. Before he goes to bed, he eats a big bowl of ice cream for calcium.
****

QUESTIONS:

1.   Determine Mr. Y's RBW (Appendix A, Tables A - 7 and 8). Please show all calculations in this report.

2.   Considering the difference between the RBW, calculate an adjusted body weight (Appendix A, Table A - 11).

3. Using the adjusted body weight in the Harris-Benedict equation and an activity factor (Appendix D, Tables D - 1 and D - 2), determine Mr. Y's daily energy needs.

4. Y's BMI and explain what the results mean (Appendix A, Tables A - 12 and 13).

5. Determine his daily protein needs (Appendix D, Table D - 2).

6. Based on Mr. Y's food record, estimate his energy and protein intake.

7. Assess Mr. Y's diet for possible nutrient deficiencies or excesses.

8. How many calories and how many grams of protein do you recommend for Mr. Y after he has completely healed from surgery?

9.   How much food will Mr. Y be able to eat at one sitting after surgery? What are the possible complications if he eats more than the recommended amount?

10.   What would be an alternative Ca source?

11.   List the foods in Mr. Y's diet that are high in fat or simple sugars and suggest an appropriate substitute.

12.   Are there any foods in Mr. Y's diet that contain substances that could bind to or react with essential nutrients and render them unavailable? Identify these foods, the harmful substance they contain, and the nutrient(s) they react with.

13.   If we chewed our food very well and ate slowly, could we lose weight? The kind and amount of food we eat is important to our health, but how and when we eat is also important. Mr. Y eats very fast with very little chewing. Discuss the importance of chewing. Include in your discussion the glucostatic theory of hunger and its possible relationship to chewing and the rate at which we eat.

14.   Considering the discussion in question 13, why was Mr. Y probably still hungry (when he sat in front of the television) after his evening meal?

15.   Mr. Y went to bed after his TV snacks and ice cream, which was not long after supper. How could this contribute to his weight gain?

16.   If Mr. Y started an exercise program, should he do aerobic exercise, anaerobic exercise, or both? Explain your answer and give examples of the types of exercises that Mr. Y could do.

17.   Describe the ways in which exercise would help Mr. Y lose weight.

18.   Define bariatric surgery.

19.   Using your textbook or your state or local diet manual, describe the principles of a gastric partitioning diet.

20. There is more than one surgical procedure for gastric partitioning. Most basic clinical texts provide illustrations of the possible procedures. Using such a text (or see the references at the end of this case), sketch at least one surgical procedure and label all anatomical parts related to the surgery.

21. What are the complications of this surgery?

22. If you had to counsel Mr. Y, what are some examples of behavior modifications you would use? Include in your discussion why Mr. Y would have to chew his food extremely well after the surgery.

23. Should Mr. Y take a vitamin and mineral supplement? Why or why not?

24. Do you think Mr. Y could still gain weight after this surgery? If so, explain how.

25.   Plan a day's menu for Mr. Y. Include in your plan the amount of food, the texture, and the time of day that it was consumed.

Related Readings

1.   Berg, F.M. (1993). *Health Risks of Obesity.* 2nd Ed. Obesity & Health, Healthy Living Institute, Hettinger, N.D.

2.   Chisholm, D.J., Samaras, K., Markovic, T., Carey, D., Lapsys, N., & Campbell, L.V. (1998). Obesity: genes, glands or gluttony? *Reprod. Fertil. Dev.* 10(1):49-53.

3.   Deitel, M. (1998). Commentary: joint pains after various intestinal bypasses and secondary to obesity. *Obes. Surg.* 8(3):265.

4.   de Witt Hamer, P.C. & Tuinebreijer, W.E. (1998). Preoperative weight gain in bariatric surgery. *Obes. Surg.* 8(3):300-301.

5.   Fobi, M.A., Chicola, K., & Lee, H. (1998). Access to the bypassed stomach after gastric bypass. *Obes. Surg.* 8(3): 289-295.

6.   Fobi, M.A., & Lee, H. (1998). The surgical technique of the Fobi-Pouch operation for obesity (the transected silastic vertical gastric bypass). *Obes. Surg.* 8(3): 283-288.

7.   Green, S.M. (1997). Obesity: prevalence, causes, health risks and treatment. *Br. J. Nurs.* 6(20):1181-1185.

8.   Hess, D.S. & Hess, D.W. (1998). Biliopancreatic diversion with a duodenal switch. *Obes. Surg.* 8(3): 267-282.

9.   Mayer, J. (1966). Some aspects of the problem of regulation of food intake and obesity. *N. Engl. J. Med.* 274(11): 610-616.

10. Mayer, J. (1996) Glucostatic mechanism of regulation of food intake. *Obes. Res.* 4(5): 493-496.

11. NHBLI (1998) *Clinical Guidelines on the Identification, Evaluation, and Treatment of Overweight and Obesity in Adults.* NHLBI Information Center, P.O. Box 30105, Bethesda, MD 20824-0105.

12. Noya, G., Cossu, M.L., Coppola, M., Tonoto, G., Angius, M.F., Fais E., & Ruggiu, M. (1998). Biliopancreatic diversion for treatment of morbid obesity: experience in 50 cases. *Obes. Surg.* 8(1): 61-66.

13. Position paper of the American Dietetic Association. (1997). Weight management. *J. Am. Diet. Assoc.* 97:71.

14. Van Itallie, TB. (1990). The glucostatic theory 1953 - 1988: roots and branches. *Int. J. Obes.* 14 Suppl 3:1-10.

15. Weight-control Information Network. November, 1993. *Understanding Adult Obesity.* NIH Publication No. 94-3680.

16. Weight-control Information Network. April, 1996. *Gastric Surgery for Severe Obesity.* NIH Publication No. 96-4006.

17.   Weight-control Information Network. December, 1996. *Prescription Medications for the Treatment of Obesity*. NIH Publication No. 96-4191.

18.   Weight-control Information Network. May, 1998. *Do you know the health risks of being overweight?* NIH Publication No. 98-4098.

19.   Whitney, E.N., C.B. Cataldo, & S.R. Rolfes. *Understanding Normal and Clinical Nutrition*, 5th Ed. West/Wadsworth. 1998.

20.   Zuidema, W.P., van Gemert, W.G., Soeters, P.B., & Greve, J.W. (1998). Pouch diverticula after banded gastroplasty for morbid obesity: report of three cases. *Obes. Surg.* 8(3):297-299.

CASE STUDY #9
OBESITY TREATED WITH VLCD

INTRODUCTION

This is the third study of obesity. The study may seem unusual study because the patient is so large, but it is based on an actual case in which the original patient weighed much more than this one. The very low-calorie weight reduction diet is not favored by nutritionist and MDs, but it is used by some for the treatment of the morbidly obese and it should be understood by health care professionals. Medical terminology is also emphasized in this case study.

SKILLS NEEDED

ABBREVIATIONS:

Knowledge of the following abbreviations is required in order to understand this case. You should learn these abbreviations before you begin to read the study.

ABGs : arterial blood gases
BIA : bioelectrical impedance
BMI : Body mass index
BMR : Basal metabolic rate
BUN : blood urea nitrogen
Ca : calcium
Cl : chloride
C/O : complains of
Cr : creatinine
d : day
dl : deciliter or 1/10 of a liter
DVT : deep vein thrombosis
ER : emergency room
g/dl : grams per deciliter
HCO_3 : bicarbonate ion
Hct : hematocrit
Hg : mercury
Hgb : hemoglobin
HPRL : prolactin
K : potassium
LBM : lean body mass
MCH : mean corpuscular hemoglobin
MCV : mean corpuscular volume
MD : medical doctor
milliIU/L : milliinternational units per liter
min : minute

mEq : milliequivalent
mg : milligram
mg/dl : milligrams per deciliter
mm : millimeter
Na : sodium
ng/dl : nanogram per liter
O_2 : oxygen
$PaCO_2$: partial pressure of carbon dioxide
PaO_2 : partial pressure of oxygen
pg : picogram
pH : a measurement of acidity or alkalinity
pt : patient
PT : physical therapy
PVC : premature ventricular contraction
qts : quarts
RBW : reference body weight
SBR : strict bed rest
sec : second
Ser Alb : serum albumin
SOB : shortness of breath
TSH : thyroid-stimulating hormone
T_3 : free triiodothyronine
T_4 : free thyroxine
VLCD : very low calorie diet
YOWM : year old white male
2° : secondary
μm^3 : cubic microns

LABORATORY VALUES:
You will need to be able to interpret the nutritional significance of the following laboratory values for this case study: BUN, Cl, Cr, glucose, Hct, Hgb, HCO_3, HPRL, K, MCH, MCV, Na, P, PaO_2, $PaCO_2$, pH, PT, Ser Alb, TSH, T_3, and T_4.

FORMULAS:
The formulas used in this case study include reference body weight, percent reference body weight, adjusted body weight (Appendix A, Tables A - 7 through 11), and body mass index (Appendix A, Tables 12 and 13). Total energy needs and activity factors can be found in Appendix D, Tables D - 1 through 3.

TB is an unemployed, 27 YOWM who used to drive a tractor-trailer rig across country. His boss fired him about two years ago when TB's weight had increased so much that it was impairing his work. During high school, TB was very active as an All-State defensive tackle. He weighed 280 lbs and was 6'3" tall. TB was heavily recruited by several universities and attended college for one year but injured his right knee and was no longer able to play football. He quit college but continued to eat as if he were still playing ball. TB held several odd jobs before he learned how to drive tractor-trailer rigs.

During high school TB ate a huge amount of food but was so active that he had no problem maintaining his weight at 280 lbs. TB's truck driving job was very sedentary but he ate more than he did in high school, mostly out of boredom. He slowly gained weight over the years until he weighed 460 lbs a little over a year ago. He became so large that he could not find a job. His friends made so much fun of him that he avoided them. TB wanted to be active in sports again but was too big to exercise and still had the knee problem. He could not walk more than 50 feet without becoming winded. Unable to fit into most cars, TB was left with nothing to do but sit at home and feel sorry for himself.

His family consisted of his mother, two younger sisters, and a younger brother. They hated to see him sitting around with nothing to do, so, with all the good intentions in the world, they chose to provide him with the one thing he loved: food. His mother and sisters would fix him anything he wanted. Eating became his favorite pastime, his escape from reality. He spent the better part of the day eating, watching TV, and gaining weight. The more weight he gained, the less he moved about, and his self-pity and depression increased. The greater the depression, the more he ate, and the cycle continued.

For the last two years TB was severely SOB. He was easily tired and even a small amount of movement caused him to be winded. Recently, he C/O throbbing headaches. His face got flushed and he perspired considerably. He had prolonged drowsiness but did not sleep well at night. Two weeks ago he started with a new symptom, he C/O pain in his left calf and was having such a hard time breathing, his sisters panicked and called 911. The paramedics gave TB some O_2 and brought him to the ER of the county hospital. He was stabilized and weighed in on the freight scales at 552 lbs. The ER physician found TB to be SOB to the point of having cyanotic extremities, particularly the toes. He was experiencing tachycardia with occasional PVCs. His left calf was tender to touch, warm, and seemed to be swollen but it was difficult to examine properly because of his massive size. The ER physician was concerned that TB had a DVT.

A venogram was completed along with blood tests and the following results were posted:

The venogram showed filling defects and diverted blood flow, positive for thrombophlebitis.

His lab values were as follows:

TEST	RESULT	NORM	TEST	RESULT	NORM	TEST	RESULT	NORM
Hgb	18.0 g/dl	14 - 18	MCH	29 pg	26 - 34	MCV	87.0 µm^3	82 - 98
Hct	50%	42 - 52	BUN	14 mg/dl	7 - 18	Cr	0.9 m/dl	0.6 - 1.2
Ser Alb	4.1 g/dl	3.9 - 5.0	Na	136 mEq/L	135 - 145	K	4.8 mEq/L	3.5 - 5.3
Glucose	140 mg/dl	65 - 110	Cl	105 mEq/L	98 - 106	P	3.0 mg/dl	2.5 - 4.5
PT	12 sec	********	Con-trol	11.4 sec	10 - 14	pH	7.31	7.35 - 7.45
PaO$_2$	88 mm Hg	80 - 90	Pa CO$_2$	58 mm Hg	35 - 45	HCO$_3^-$	22 mEq/L	24 - 28

The ER physician diagnosed the patient as follows:
- ▸ 1. Morbidly obese
- ▸ 2. Pickwickian syndrome
- ▸ 3. Mild tachycardia 2° to obesity/Pickwickian syndrome
- ▸ 4. Mild respiratory acidosis 2° to Pickwickian syndrome
- ▸ 5. DVT 2° to obesity and immobilization
- ▸ 6. Polycythemia 2° to Pickwickian syndrome

His treatment included:
- ■ 1. SBR with heat applied to left calf
- ■ 2. Heparin therapy
- ■ 3. Head of bed elevated
- ■ 4. VLCD of 700 kcals

The VLCD is sometimes used with the very obese. Several VLCD plans are available that are sponsored by hospitals, wellness centers, etc. The patients are usually closely monitored by an MD, RD, exercise physiologist, and psychologist. Most plans provide 70 g of protein, 100 g of carbohydrate, and about 800 kcals. The usual range is from 600 to 900 kcals. The entire diet is usually in a liquid form for approximately the first three months. Emphasis is placed on maintaining a sodium restriction (2 g of sodium per day), ingesting no more than 300 mg of caffeine per day, consuming at least 2 qts of fluid per day, incorporating an individualized exercise prescription, and attending behavior modification classes. The program also includes a bi-weekly visit with a physician and bi-weekly analyses of blood and urine. This continues for approximately 12 weeks. Solid food is gradually added to the diet, replacing the liquid supplement until about 1000 to 1200 kcals are reached. A registered dietitian teaches weekly nutrition classes and supervises the addition of solid food to the diet. A diet of solid food without supplements is accomplished by the 5th or 6th week after the 12 week fast. A high potency multivitamin and mineral tablet along with 50 mEq K and 800 mg of Ca/d are recommended during the fasting period.
****

QUESTIONS:

1. The events leading up to TB's depression and excessive weight gain are not uncommon. When someone who leads an active lifestyle slows down suddenly, rapid weight gain will occur. List the events that contributed to TB's problem and explain how each of these events contributed.

2. If you had the opportunity to counsel TB about his weight when he weighed 480 lbs, describe how you would handle this consultation. There is no single correct answer to this question. The following are some examples of the problems you may want to address.

   A. A person of this size is used to consuming huge amounts of food. What could TB do to help keep his mind off of food?
   B. Think about an exercise program for TB. What could he do for exercise?
   C. Determine if there are any support groups for the obese in your area. If so, describe how to involve TB with that support group.

3. Calculate TB's BMI and evaluate the results (Appendix A, Tables A - 12 and 13).

4. Determine TB's RBW and percent of RBW (Appendix A, Tables A - 7 through 10). Please show all calculations.

5.   There is considerable variation between TB's RBW and actual body weight. Calculate TB's adjusted body weight (Appendix A, Table A - 11).

****

Information Box 9 - 1

Question 7 is a difficult one. You will not find an answer in any resource that accurately addresses all sides of the issue. Patients of this size are difficult to evaluate. The caloric/protein requirements will depend on a number of factors, including age, level of activity, hormone levels, disease condition, the genetic influences on basal metabolic rate, and body composition. Body composition, or ratio of lean body mass (LBM) to adipose tissue, is an important consideration but is difficult to evaluate with a patient this large. BIA is only accurate when a patient's hydration status is normal. A "normal" hydration status of an obese person that is dieting is difficult to evaluate. Body composition can be determined by means of skinfold measurements, but the calipers that exist are too small to make this test possible. The patient would be too large to fit into the equipment for underwater weighing. The best answer should come from a dietitian who is experienced in working with the morbidly obese. (☞ *Hint: Evaluate the patient as well as you can based on the information provided and find a RD with experience working with the morbidly obese to obtain an "oral communication" reference.*)

6.   In this case, do you think that using the adjusted body weight is an accurate way to calculate TB's caloric and protein needs? Explain your answer. If you do not think it is accurate, relate how you estimate TB's caloric and protein needs.

7.   Calculate TB's protein requirement (Appendix D, Table D - 2).

8.   Explain the rationale behind the VLCD.

9.   Describe the drawbacks and potential hazards of the VLCD. Do the benefits outweigh the disadvantages in this case?

10.   Do you think that the usual amount of kcals and protein allowed on a VLCD would be adequate in this case?   Discuss your answer and indicate the amount of kcals and protein you would allow.

11.   Why would caffeine be restricted?

12.   Briefly discuss why most weight reduction diets emphasize restricting sodium.

13.   What relationship, if any, does potassium have with this diet?

14.   Discuss the importance of behavior modification.

15.   With a patient of this size, how could the exercise component be implemented?

16.   What symptoms can be expected during the first several days of such a fast?

17.   When the fast is over the patient slowly returns to a regular diet. What is the rationale for this?

18.   Do you agree with the prescription of a 1000 to 1200 kcal diet for this patient after the fast is over? Discuss your answer and if you do not agree, what kcal level do you recommend?

19. Plan a 1100 kcal diet for a patient weighing 562 lbs.

20. Review the first table of lab values. Describe the relationship, if any, between the abnormal lab values and TB's nutritional status.

21. Define the following terms:

Cyanotic:

Tachycardia:

Thrombophlebitis:

Morbid obesity:

Pickwickian syndrome:

Respiratory acidosis:

Polycythemia:

Decubiti:

Prone:

22.   Would TB be a candidate for gastric surgery to prevent obesity? Explain your answer.

Related Readings

1. Berg, F.M. (1993). Health Risks of Obesity. 2nd Ed. Obesity & Health, Healthy Living Institute, Hettinger, N.D.

2. Chisholm, D.J., Samaras, K., Markovic, T., Carey, D., Lapsys, N., & Campbell, L.V. (1998). Obesity: genes, glands or gluttony? *Reprod. Fertil. Dev.* 10(1):49-53.

3. Fischbach, F.T. (1995). *A Manual of Laboratory & Diagnostic Tests.* 5th Ed. Philadelphia. J.B. Lippincott Company.

4. Grant, A. & DeHoog, S. (1991). *Nutritional Assessment and Support.* 4th Ed. Seattle, Washington. A. Grant and S. DeHoog, publishers.

5. Green, S.M. (1997). Obesity: prevalence, causes, health risks and treatment. *Br. J. Nurs.* 6(20):1181-1185.

6. Lee, R.D. & Nieman, D.C. (1993). Nutritional Assessment. Brown and Benchmark, pp. 180-188.

7. Mayer, J. (1996). Some aspects of the problem of regulation of food intake and obesity. *N. Engl. J. Med.* 274(11): 610-616.

8. Mayer, J. (1996) Glucostatic mechanism of regulation of food intake. *Obes. Res.* 4(5): 493-496.

9. NHBLI (1998) *Clinical Guidelines on the Identification, Evaluation, and Treatment of Overweight and Obesity in Adults.* NHLBI Information Center, P.O. Box 30105, Bethesda, MD 20824-0105.

10. Position paper of the American Dietetic Association. (1997). Weight management. *J. Am. Diet. Assoc.* 97:71.

11. Weight-Control Information Network. November, 1993. *Understanding Adult Obesity.* NIH Publication No. 94-3680.

12. Weight-Control Information Network. March, 1995. *Weight Cycling.* NIH Publication No. 95-3901.

13. Weight-Control Information Network. March, 1995. *Very Low-Calorie Diets.* NIH Publication No. 95-3894.

14. Weight-Control Information Network. December, 1996. *Prescription Medications for the Treatment of Obesity.* NIH Publication No. 96-4191.

15. Weight-control Information Network. May, 1998. *Do you know the health risks of being overweight?* NIH Publication No. 98-4098.

16.   Weinsier, R.L., Hunter, G.R., Heini, A.F., Goran, M.I., & Sell, S.M. (1998). The etiology of obesity: relative contribution of metabolic factors, diet, and physical activity. *Am. J. Med.* 105(2):145-150.

17.   Whitney, E.N., Cataldo, C.B., & Rolfes, S.R. (1998). *Understanding Normal and Clinical Nutrition*, 5th Ed. West/Wadsworth.

CASE STUDY #10
WEIGHT MANAGEMENT WITH CULTURAL CONCERNS

INTRODUCTION
This case study is about a migrant worker from Mexico who is overweight and has additional problems as well. Some cultural practices of Mexico are introduced. A review of the cultural and dietary practices of Mexican-Americans would be beneficial to the understanding of this case.

SKILLS NEEDED

ABBREVIATIONS:
Knowledge of the following abbreviations is required in order to understand this case. You should learn these abbreviations before you begin to read the study.

bid : twice a day
BMI : body mass index
BUN : blood urea nitrogen
cc/h : cubic centimeters per hour
Cl : chloride
Cr : creatinine
D_5W : dextrose 5% in water
ER : emergency room
GI : gastrointestinal
g/dl : grams per deciliter
Hct : hematocrit
Hgb : hemoglobin
I.V. : intravenous
K : potassium
MCV : mean corpuscular volume

mEq/L : milliequivalent per liter
mg : milligram
mg/dl : milligram per deciliter
Na : sodium
NPO : nothing by mouth
P : phosphorous
po : by mouth
qid : four times a day
q4h : every four hours
RBW : reference body weight
Ser Alb : serum albumin
Trig : triglycerides
x1d : for one day
μm^3 : cubic microns
¼NS : one fourth strength normal saline

LABORATORY VALUES:
You will need to be able to interpret the nutritional significance of the following laboratory values for this case study: BUN, Cr, glucose, Hct, Hgb, MCV, Na, K, Cl, P, Trig, and Ser Alb.

FORMULAS:
The formulas used in this case study are reference body weight and percent reference body weight (Appendix A, Tables A - 7 through A - 10).

MEDICATIONS:
Become familiar with the following medications before reading the case study. Note the diet-drug interactions, dosages and methods of administration, gastrointestinal tract reactions, etc.

1. Phenergan (promethazine hydrochloride); 2. Lomotil (diphenoxylate hydrochloride).

Mrs. MG is a 43 year old Mexican who has been living in this country for several years. She and part of her family migrated to Texas from rural Mexico looking for work. Mrs. MG's household includes her husband, five children, and her mother-in-law. Mrs. MG and her husband have been working as migrant farm workers for less than minimum wage. Mrs. MG has been healthy but recently has been experiencing symptoms of gastroenteritis, including bloating, nausea, and diarrhea. She vomits occasionally but not on a regular basis. She has not been able to eat very much because of her illness. This continued for almost a week before her husband decided to bring her to a doctor. By the time she went to the emergency room, she was very weak, dehydrated, confused, and disoriented. Observing her condition and her obvious state of poverty, the ER physician decided to admit her for a couple of days so that she could recover from what he thought was simple dehydration and a stomach virus.

An examination of Mrs. MG revealed the following: Mrs. MG was 5'3" and weighed 165 lbs. She has always been overweight but has slowly been losing weight over the last year. Her labs completed in the ER were as follows:

TEST	RESULT	NORM	TEST	RESULT	NORM	TEST	RESULT	NORM
Hgb	17.0 g/dl	12 - 16	Trig	250 mg/dl	40 - 160	MCV	79.0 μm^3	82 - 98
Hct	50%	36 - 48	BUN	29 mg/dl	7 - 18	Cr	1.8 mg/dl	0.6 - 1.3
Ser Alb	5.1 g/dl	3.9 - 5.0	Na	150 mEq/L	135 - 145	K	5.1 mEq/L	3.5 - 5.3
Glucose	220 mg/dl	65 - 110	Cl	108 mEq/L	98 - 106	P	3.8 mg/dl	2.5 - 4.5

The physician admitted Mrs. MG with the following orders:

Doctor's Orders

10/1/98 Admit Mrs. MG to gen med service
I.V. - D5 W 1/2 NS @ 100 cc/h
Phenergan 25 mg PO q 4h
Lomotil 5mg PO qid x 1d then 5mg bid
BRP as able
NPO
RD to see pt.
[illegible]

The dietitian was new to the south Texas area and was not yet familiar with Mexican-American customs. Mrs. MG and her husband spoke very poor English, and the RD did not speak Spanish. She had a very difficult time communicating and could not get Mrs. MG to commit to anything she asked her to do. The RD noticed that Mrs. MG looked at her husband to see his reaction to everything the dietitian said. The RD decided to end the conversation in order to go do some research and obtain the help of an interpreter.

Through her research on Mexican culture, the RD learned that the family and the extended family are of central importance to an individual's life. Most decisions are made by the male head of the household, who is strongly influenced by his mother. Individual members of the family do not like to make hasty decisions but prefer to talk things over with the family first. This is why the RD could not get Mrs. MG to commit to anything. She also learned that Mrs. MG's culture relies on home remedies first before seeking outside help. When they do seek outside help, it is usually from a folk healer or native healer rather than from a physician. This explains why Mrs. MG's husband waited so long before he took her to the hospital. Late that afternoon the RD went back to see Mrs. MG with a Mexican diet technician who understood the culture and could speak Spanish. Before going to see Mrs. MG again, the RD checked the chart to see if anything new had been entered. She found a new set of lab values that had been completed that afternoon. The labs pertinent to the patient's nutritional status were as follows:

TEST	RESULT	NORM	TEST	RESULT	NORM	TEST	RESULT	NORM
Hgb	12.0 g/dl	12 - 16	Trig	210 mg/dl	40 - 160	MCV	77.0 µm³	82 - 98
Hct	33%	37 - 47	BUN	16 mg/dl	7 - 18	Cr	1.1 mg/dl	0.6 - 1.3
Ser Alb	4.3 g/dl	3.9 - 5.0	Na	143 mEq/L	135 - 145	K	3.8 mEq/L	3.5 - 5.3
Glucose	200 mg/dl	65 - 110	Cl	105 mEq/L	98 - 106	P	2.6 mg/dl	2.5 - 4.5

****

QUESTIONS:
1.   What is Mrs. MG's RBW and percent RBW (Appendix A, Tables A - 7 through 10)?

2.   Calculate Mrs. MG's BMI (Appendix A, Tables A -12 and 13).

3.　　Write out the physician's orders in long-hand and explain abbreviations as appropriate.

4.　　After hydration with I.V. fluids, several lab values changed. Which values changed with the addition of the fluid and why?

5.　　Do any of the results in the first set of lab values indicate anemia? If so, identify the lab value(s) and the type of anemia.

6.　　Do any of the results in the second set of lab values indicate anemia? If so, identify the lab value(s) and the type of anemia.

Information Box 7 - 1

The number of calories provided by D_5W will contribute to the patient's caloric intake and certainly needs to be accounted for. However, it must be remembered that this intake will never be sufficient to meet anyone's caloric needs. D_5W is equivalent to half-strength Kool-Aid. To determine the grams of glucose provided by I.V. infusions, determine the amount of cc infused per day (flow rate X 24 hrs) and multiply by the percent glucose in the solution. In this case it is 5%. To determine the caloric contribution from this, multiply the grams by 3.4. Glucose is normally 4 kcals per gram, but in solution it is in a hydrated state and is 3.4 kcals per gram. [1]

[1] Rombeau, J.L., and Caldwell, M.D. (1993). Clinical Nutrition: PARENTERAL NUTRITION, 2nd
     Ed. W.B. Saunders Company, Philadelphia pg. 310.

---

7.   Mrs. MG received D_5W in 1/4 NS at 100 cc/hr. How many grams of glucose and how many calories would that provide per day?

****

The RD went into Mrs. MG's room with the Mexican diet tech and made all of the appropriate introductions. Mrs. MG's husband was there so the RD interviewed both of them with the help of the interpreter. The results were as follows:

Mrs. MG and her family, with the guidance of a local folk healer, thought she had "Empacho." This is something similar to gastroenteritis and is believed to be caused by a blockage in the GI tract. The RD questioned her about symptoms of frequent urination, thirst, and hunger and learned that Mrs. MG experienced all of these in the past few months. The RD also found out that Mrs. MG's father had a disease that caused him to lose his foot, but she did not know what it was.

Her diet in this country had not changed much from her native diet in Mexico. Every day she ate some kind of beans, pinto and black beans being her favorites. Tortillas were also part of the everyday diet. She ate fried corn tortillas. Pork was the favorite meat, particularly "chorizos," if it was available. Fried foods were high on her list, typically lard was used as the fat. Vegetables included tomatoes, onions, squash, garlic, and chili peppers of all kinds. She did not drink very much milk; lactose intolerance seemed to be a problem. Mrs. MG liked cheese and fresh fruit, but both were rarely included in the diet because of the cost. Foods that she has enjoyed since she has been in this country are hamburgers, french fries, potato chips, sodas, and candy.
****

QUESTIONS CONTINUED:
8.   Considering Mrs. MG's family medical history, weight history, and most recent set of lab values, what was the RD looking for with her line of questions about frequent urination, thirst, hunger, etc.?

9.   List the diet principles Mrs. MG should follow and, based on the principles you list, prescribe an appropriate specific diet for Mrs. MG.

10.   In view of the answer to question 8, list the foods in Mrs. MG's diet that need to be restricted and give a reason why.

11.   Based on Mrs. MG's recall, what nutrients are apparently deficient in her diet?

12.   List any protein combinations in MG's diet that make complete proteins.

13.   Considering her income, education, and cultural practices, outline a nutrition care plan you would use for Mrs. MG. Include the teaching techniques you would use.

14. Define empacho and chorizos.

Related Readings

1. Algert, S.J., Brzezinski, E., & Ellison, T.H. (1998). *Ethnic and Regional Food Practices, A Series: Mexican American Food Practices, Customs, and Holidays.* The American Dietetic Association, Chicago, IL and the American Diabetes Association, Alexandria, VA.

2. Berg, F.M. (1993). Health Risks of Obesity. 2nd Ed. Obesity & Health, Healthy Living Institute, Hettinger, N.D.

3. Chisholm, D.J., Samaras, K., Markovic, T., Carey, D., Lapsys, N., & Campbell, L.V. (1998). Obesity: genes, glands or gluttony? *Reprod. Fertil. Dev.* 10(1):49-53.

4. Fischbach, F.T. (1995). *A Manual of Laboratory & Diagnostic Tests.* 5th Ed. Philadelphia. J.B. Lippincott Company.

5. Green, S.M. (1997). Obesity: prevalence, causes, health risks and treatment. *Br. J. Nurs.* 6(20):1181-1185.

6. Grant, A. & DeHoog, S. (1991). *Nutritional Assessment and Support.* 4th Ed. Seattle, Washington. A. Grant and S. DeHoog, Publishers.

7. Mayer, J. (1966). Some aspects of the problem of regulation of food intake and obesity. *N. Engl. J. Med.* 274(11): 610-616.

8. NHBLI (1998) *Clinical Guidelines on the Identification, Evaluation, and Treatment of Overweight and Obesity in Adults.* NHLBI Information Center, P.O. Box 30105, Bethesda, MD 20824-0105.

9. Position paper of the American Dietetic Association. (1997). Weight management. *J. Am. Diet. Assoc.* 97:71.

10. Rombeau, J.L., and Caldwell, M.D. (1993). Clinical Nutrition: PARENTERAL NUTRITION, 2nd Ed. W.B. Saunders Company, Philadelphia.

11. Weight-Control Information Network. November, 1993. *Understanding Adult Obesity.* NIH Publication No. 94-3680.

12. Weight-Control Information Network. March, 1995. *Weight Cycling.* NIH Publication No. 95-3901.

13. Weight-Control Information Network. May, 1998. *Do you know the health risks of being overweight?* NIH Publication No. 98-4098.

14. Weinsier, R.L., Hunter, G.R., Heini, A.F., Goran, M.I., & Sell, S.M. (1998). The etiology of obesity: relative contribution of metabolic factors, diet, and physical activity. *Am. J. Med.* 105(2):145-150.

15. Whitney, E.N., Cataldo, C.B., & Rolfes, S.R. (1998). *Understanding Normal and Clinical Nutrition*, 5th Ed. West/Wadsworth.

CASE STUDY #11
GASTRITIS IN AN ELDERLY WOMAN

INTRODUCTION
This is part one of a two-part case study involving a typical scenario of aging, weight gain, and the medical complications that accompany this process. It is concerned with weight gain, gastritis, and esophageal reflux, complicated by medications for various other disorders.

SKILLS NEEDED

ABBREVIATIONS:
Knowledge of the following abbreviations is required in order to understand this case. You should learn these abbreviations before you begin to read the study.

ac : before meals	po : by mouth
bid : twice a day	prn : as needed
BMI : body mass index	qd : every day
D/C : discontinue	qid : four times a day
EGD : esophagogastroduodenoscopy	RBW : reference body weight
g : gram	UGI : upper gastrointestinal
hs : hour of sleep	YOWF : year old white female
mg : milligram	2xd : two times daily

FORMULAS:
The formulas used in this case study include reference body weight, adjusted body weight, percent reference body weight, and body mass index (Appendix A, Tables A - 7 through 13).

MEDICATIONS:
Become familiar with the following medications before reading the case study. Note the diet-drug interactions, dosages and methods of administration, gastrointestinal tract reactions, etc.

1. Tenormin (atenolol); 2. Lasix (furosemide); 3. Tums (calcium carbonate); 4. Aspirin (acetylsalicylic acid); 5. Clinoril (sulindac); 6. Harmonyl (deserpidine); 7. Carafate (sucralfate); 8. Zantac (ranitidine hydrochloride);

Mrs. CL is a 79 YOWF with a history of indigestion that goes back many years. For as long as she can remember she has been taking Tums after most of her meals, particularly the evening meal. Her indigestion worsened over the years and her ingestion of Tums increased proportionately. By the time she was 77, she was taking Tums after every meal and before going to bed at night. The increase was so gradual that Mrs. CL did not realize how many Tums she was taking. She was using a generic brand of calcium carbonate in a very large bottle that she bought in a discount store. This practice had become such an accepted part of her routine that by her 79th birthday it was as natural for her to take a Tums after a meal as it was for her to wipe her mouth with a napkin.

Mrs. CL's weight and age increased as her intake of Tums increased, while her activity level progressively decreased. At age 79, Mrs. CL was 5'1" and weighed 175 lbs. When she was married 49 years ago, she weighed 110 lbs. Most of her weight gain occurred after age 40, but the most dramatic increase occurred after Mrs. CL retired at age 63. Her weight history can be found in Table I. Mrs. CL was an excellent cook and restaurant manager for 10 years prior to her retirement. After her retirement, she helped keep herself busy by cooking for large groups of people at church and civic functions, family gatherings, etc. At some point she developed the habit of tasting everything several times while it was cooking. This turned into almost constant eating while cooking.

TABLE I	
AGE	WEIGHT
30	110
40	128
63	148

As active as Mrs. CL was after retirement, she was not expending nearly as much energy as she did prior to retirement. The lack of exercise, increased eating, and advancing age contributed to the dramatic weight gain. As Mrs. CL aged, her ability to remember and reason diminished. She is on a fixed income, relying on social security, and she lives with her son and daughter-in-law. They do not charge her rent, but she takes care of her own medical bills and buys some of the food. She feels like she has to contribute to the household by cooking for the family. They allow this and enjoy her cooking most of the time, but they have noticed that as she ages, she is beginning to lose her ability to modify recipes and ad lib in the kitchen. She started cooking the same meals very frequently and preparing unbalanced meals, such as rice, potatoes, and peas with bread. They prepared a three-week cycle menu for her to follow for variety and balance.

Mrs. CL's family did not realize how many Tums she was taking. She never discussed the "burning in her chest" with them. They encouraged her on a regular basis to lose weight without success. Additional factors that affected Mrs. CL's health included arthritis that worsened as she grew older. She took Clinoril (sulindac), 150 mg bid for arthritis. She found that aspirin (acetylsalicylic acid) did her as much good, was cheaper, and could be taken more often. Her aspirin intake also began to increase over the years and by the time she was 79 years old it was another one of her accepted "natural habits." Mrs. CL also had a problem with hypertension. Her blood pressure was usually about 160/94 without medication and 140/84 with medication. She was taking Harmonyl (deserpidine) for her hypertension, 0.25 mg po qd. She was also taking Lasix (furosemide), 20 mg prn for swelling in her feet.

One evening at supper, Mrs. CL started to panic and indicated to her son and daughter-in-law that she was choking. After her son performed the Heimlich maneuver on her, she coughed up a piece of chicken. She said she tried to swallow and could not. The chicken just got stuck in her throat and would not go down. She was afraid to eat after that and wanted to go to the doctor. It was at this time that she told her son of the burning she had been experiencing in her chest. She said that it had been getting worse over the past few weeks. He made an appointment for her with the family doctor, who sent her to a gastroenterologist. He performed a UGI series and an esophagogastroduodenoscopy (EGD) on her. The results indicated gastritis without ulceration and decreased esophageal motility with reflux esophagitis. He told her to D/C aspirin, Clinoril, and Harmonbyl and to see her family physician for replacement drugs. He also prescribed Carafate (sucralfate), 1 g po qid 1 hour ac and hs, and Zantac (ranitidine hydrochloride), 150 mg bid. His prognosis was good for the gastritis, but said that the esophageal reflux would probably give her problems for the rest of her life and she would have to learn to live with it. He told her to follow a soft bland diet.

Mrs. CL went to her family physician and he changed her hypertensive medication to Tenormin (atenolol), 50 mg once qd; he left her arthritis medication as is and told her not to take it for more than ten days and then wait at least a week before restarting it. He also told her to take it only when she was having continuous severe pain. He did not change the Lasix (furosemide).

Mrs. CL followed her physician's orders and took all of the medications until they ran out, which took about one month. She did not get them refilled because she was feeling better and they were too expensive. The Carafate was about $1.00 per tablet. She had Medicare and supplemental insurance, but neither paid for her medications.

****

QUESTIONS:

1.   What is Mrs. CL's reference body weight (Appendix A, Tables A - 7 through 9).

2.   Calculate her adjusted body weight (Appendix A, Table A - 11).

3.   Calculate her BMI (Appendix A, Tables A - 12 and 13).

4.   Discuss the effects of age and exercise on basal metabolic rate.

5.   Considering Mrs. CL's age and physical condition, can you recommend an exercise routine that would be appropriate? Explain your answer.

6.   Discuss the possible reasons for Mrs. CL's gastritis.

7.   List the dietary principles to be followed for esophageal reflux.

8.   Define:

arthritis:

UGI series:

Hypertension:

Reflux esophagitis:

Gastroenterologist:

Gastritis without ulceration:

Decreased esophageal motility:

Esophagogastroduodenoscopy:

9.   Describe the action and side effects of Clinoril.

10.   What is considered to be normal and acceptable blood pressure?

11.   Describe the action and side effects of:

Harmonyl:

Lasix:

Carafate:

Zantac:

Tenormin:

12.   Why do you think the physician did not change the Clinoril? (☞ *Hint: look up the other drugs in its class, i.e., the nonsteroidal anti-inflammatory drugs and compare to Clinoril*).

13.   The physician did not tell Mrs. CL much about her diet. Expound on what he said by writing a nutritional care plan to fit Mrs. CL's nutritional needs and lifestyle.

14.   Failing to refill prescriptions because of the cost is typical for the elderly who are on fixed incomes. Discuss the limitations that financial restrictions can place on this population, and offer some suggestions for techniques that the health care team can use to emphasize the importance of the medications.

15. Some areas have assistance programs for individuals who cannot afford the medications they need. Determine if such a program exists in your area. If so, discuss the charac teristics and value of the program. (☞ *Hint: a hospital social worker is a good source of information for this).*

Related Reading

1.   Chernoff, R. (1991). *Geriatric Nutrition.* 3rd Ed. Gaithersburg, Maryland. Aspen Publishers.

2.   Blumberg, J. (1997). Nutritional needs of seniors. *J. Am. Coll. Nutr.* 16(6):517-523.

3.   Bozymski, E.M. & Isaacs, K.L. (1991). Special diagnostic and therapeutic considerations in elderly patients with upper gastrointestinal disease. *J. Clin. Gastroenterol.* 13(S2):S65-S75.

4.   Fischbach, F.T. (1995). *A Manual of Laboratory & Diagnostic Tests.* 5th Ed. Philadelphia. J.B. Lippincott Company.

5.   Koskenpato, J., Kairemo, K., Korppi-Tommola, T. & Farkkila, M. (1998). Role of gastric emptying in functional dyspepsia: a scintigraphic study of 94 subjects. *Dig. Dis. Sci.* 43(6):1154-1158.

6.   Lovat, L.B. (1996). Age related changes in gut physiology and nutritional status. *Gut.* 38(3):306-309.

7.   Nutrition Screening Initiative. (1991). A project of the American Academy of Family Physicians, the American Dietetic Association, and the National Council on Aging, funded in part by Ross Laboratories. 2626 Pennsylvania Ave., NW, Suite 301, Washington, DC 20037.

8.   Perlman, P.E. & Adams, W. (1989). Physiologic changes as patients get older. *Postgrad Med.* 85(2):213-214.

9.   Position paper of the American Dietetic Association. (1996): Nutrition, aging, and the continuum of care. *J. Am. Diet. Assoc.* 96:1048.

10.   Position paper of the American Dietetic Association. (1998): Liberalized diets for older adults in long-term care. *J. Am. Diet. Assoc.* 98:201.

11.   Prendergast, A. & Fulton, F.L. (1997). *Medical Terminology: A Text/Workbook.* 4th Ed. Redwood City, California. Addison-Wesley Nursing.

12.   Pronsky, Z.M. & Solomon, E. (1997). *Food-Medication Interactions.* 10th Ed. Phoenix, Arizona. Food-Medication Interactions, Publishers and Distributors.

13.   Saltzman, J.R. & Russell, R.M. (1998). The aging gut. Nutritional issues. *Gastroenterol Clin. North Am.* 27(2):309-324.

14.   Schlenker, E.D. (1997). *Nutrition in Aging.* St. Louis, Missouri. Mosby - Year Book, Inc.

15.   Whitney, E.N., Cataldo, C.B., & Rolfes, S.R. (1998). *Understanding Normal and Clinical Nutrition,* 5th Ed. West/Wadsworth.

CASE STUDY #12
CHEST PAIN IN AN ELDERLY WOMAN

INTRODUCTION
This is part two of a two-part case study involving a typical scenario of aging, weight gain, and the medical complications that accompany this process. It is concerned with gastritis and esophageal reflux, complicated by medications for various other disorders. The patient's symptoms were alleviated in part one, but the patient did not follow up with her medications and now requires additional treatment.

SKILLS NEEDED

ABBREVIATIONS:
Knowledge of the following abbreviations is required in order to understand this case. You should learn these abbreviations before you begin to read the study.

ac : before meals
bid : twice a day
C/O : complains of
EKG : electrocardiogram
g : gram
hs : hour of sleep (at bed time)
I.V. : intravenous

mci : millicuries
mg : milligram
MNT : medical nutrition therapy
po : by mouth
qid : four times a day
R/O : rule out
YOWF : year old white female

FORMULAS:
The formulas used in this case study include reference body weight, adjusted body weight, Harris-Benedict equation, and activity factors (Appendix A, Table A - 7 through 11 and Appendix D, Tables D - 1, 2, and 5).

MEDICATIONS:
Become familiar with the following medications before reading the case study. Note the diet-drug interactions, dosages and methods of administration, gastrointestinal tract reactions, etc.

1. Tylenol (acetaminophen); 2. Technesium Cardiolite; 3. Tums (calcium carbonate); 4. Aspirin (acetylsalicylic acid); 5. Clinoril (sulindac); 6. Persantine (dipyridamole); 7. Carafate (sucralfate); 8. Zantac (ranitidine hydrochloride).

Mrs. CL followed her bland diet and was careful to chew everything well before swallowing. She also took her medication as prescribed. The swallowing difficulty seemed to go away, and the burning stopped. When the medication ran out, Mrs. CL did not refill her prescription because the medicine was so expensive. She thought that she would wait and see what happened. Several weeks went by and the symptoms did not return. Now and then there would be a little burning and she would take two Tums, but she was used to that and did not consider it to be unusual. Both physicians, with the help of her son, convinced her not to take any more aspirins. She tried Tylenol (acetaminophen) for her arthritis pain, but it did not help as much so, as time went on, she began taking the Clinoril (sulindac) more frequently. She also began to take some aspirin again. She was told that if she took too much of the Clinoril, since it was stronger than aspirin, it could cause harm to her stomach. She figured that if Clinoril was stronger than aspirin, it made sense not to take it as much and take aspirin instead. As time passed, she ended up taking Tums, aspirin, and Clinoril as frequently as she had before.

Two years have passed since the previous incident and Mrs. CL is now an 81 YOWF with the same history of indigestion (see Case Study #7). She is older, more forgetful, and a little heavier than she was two years ago. Her height is 5'1" and she now weighs 185 lbs. Her son and daughter-in-law have cautioned her about her weight, but there is not much they can do to control her eating during the day when they are working.

Every Friday, her daughter-in-law is off work and is usually at home. One Friday, Mrs. CL came into the kitchen, barely able to walk or breath. She C/O a severe pain in the center of her chest. Her eyes were dilated, and she appeared to be very frightful. The daughter-in-law, knowing her history, sat her down and calmed her while questioning her. She found out that the pain did not radiate to her shoulder or arm but did go up into her neck. She was not clammy to touch and was not diaphoretic. She did not experience a feeling of heavy pressure pushing in on her chest. Was it a heart attack or gastric reflux? Considering her history, age, weight, hypertension, etc., there was no way to be sure. The daughter-in-law called her husband and the physician. The physician told her to bring Mrs. CL in to see him right away. By the time she arrived at the doctor's office, the pain was gone and she was calm. He did blood work to look at enzymes and did an EKG. All were negative. After she calmed down, she told her son that the pain was similar to the reflux pain she had experienced in the past, but more severe. It seemed she had a severe case of reflux esophagitis that went up into her neck.

To R/O a heart blockage, the physician decided to test further and scheduled her for a Cardiolite Imaging. This procedure is performed in the Department of Nuclear Cardiology. Mrs. CL reported at 8:00 A.M. and after receiving the appropriate instructions about the risks of the test, signed a consent form. She was then injected I.V. with 8 mci of Technesium Cardiolite. This is a radioactive isotope that enables a gamma camera to record images of the heart. After the injection, there was a minimum waiting period of one hour before the patient was placed on a special table, and the heart was scanned with a gamma camera for 16 minutes. The results were recorded in a computer and could be viewed from a computer screen. At this point, the patient is either exercised physically (on a treadmill) or chemically (I.V. injection) and the heart imaged again to determine if there are any differences between resting perfusion and exercise perfusion. Mrs. CL was not able to walk on a treadmill, so she had to be injected I.V. with Persantine (amount depends on patient's weight). This stresses the heart and its effects are equivalent to those of exercise. The patient is then injected with 22 mci of Technesium Cardiolite. After a minimum waiting period of 30 minutes, the patient is again placed on the special table and a gamma camera takes images of the heart for 16 minutes.

The results of this test were also negative, and it was concluded that Mrs. CL experienced severe pain from gastritis and reflux esophagitis. Her physician ordered her to take Carafate (sucralfate), 1 g po qid 1 hour ac and hs as before, but changed Zantac to Axid (nizatidine), 150 mg bid. He told her to continue Clinoril (sulindac) 150 mg bid but not to take it for more than 10 days, and to wait two weeks before resuming the medication. He also told her to avoid aspirin entirely. He gave her a soft bland diet to follow and this time made her an appointment with the clinic dietitian.
****

QUESTIONS:

1.   Define:

   Diaphoretic:

   Gastric Reflux:

   Reflux Esophagitis:

   Radioactive Isotope:

   Gamma Imaging:

   Perfusion:

   Gastritis:

2.   Calculate Mrs. CL's BMI (Appendix A, Tables A - 12 and 13).

3.   What is Mrs. CL's reference body weight (Appendix A, Table A - 7 through 9)?

4.   Calculate Mrs. CL's adjusted body weight (Appendix A, Table A - 10).

5.   With the above information, calculate Mrs. CL's total kcal need using the Harris-Benedict equation and an appropriate activity factor (Appendix D, Tables D - 1, 2, and 5).

6.   List the principles of the diet Mrs. CL should be on.

7.   Define Medical Nutrition Therapy.

8.   What MNT would you recommend for Mrs. CL?

9.   Compare acetylsalicylic acid and acetaminophen as to action and nutritional complications.

10. Explain how the actions of the drugs Mrs. CL was taking could have caused her problem, particularly with her medical history.

11. Describe the action and nutritional implications of Axid.

12. Axid and Carafate both cost about $1.00 per pill. Add up what this will cost Mrs. CL per day and per month. Mrs. CL is on a fixed income. Discuss what you would do if you had to counsel this patient, knowing that she is going to tell you that she cannot afford the medications.

13. Discuss any psychosocial factors that could affect Mrs. CL.

14. Describe the actions of Technesium Cardiolite and Persantine.

Related References

1. Chernoff, R. (1991). *Geriatric Nutrition.* 3rd Ed. Gaithersburg, Maryland. Aspen Publishers.

2. Blumberg, J. (1997). Nutritional needs of seniors. *J. Am. Coll. Nutr.* 16(6):517-523.

3. Bozymski, E.M. & Isaacs, K.L. (1991). Special diagnostic and therapeutic considerations in elderly patients with upper gastrointestinal disease. *J. Clin. Gastroenterol.* 13(S2):S65-S75.

4. Fischbach, F.T. (1995). *A Manual of Laboratory & Diagnostic Tests.* 5th Ed. Philadelphia. J.B. Lippincott Company.

5. Koskenpato, J., Kairemo, K., Korppi-Tommola, T. & Farkkila, M. (1998). Role of gastric emptying in functional dyspepsia: a scintigraphic study of 94 subjects. *Dig. Dis. Sci.* 43(6):1154-1158.

6. Lovat, L.B. (1996). Age related changes in gut physiology and nutritional status. *Gut.* 38(3):306-309.

7. Nutrition Screening Initiative. (1991). A project of the American Academy of Family Physicians, the American Dietetic Association, and the National Council on Aging, funded in part by Ross Laboratories. 2626 Pennsylvania Ave., NW, Suite 301, Washington, DC 20037.

8. Perlman, P.E. & Adams, W. (1989). Physiologic changes as patients get older. *Postgrad Med.* 85(2):213-214.

9. Position paper of the American Dietetic Association. (1996): Nutrition, aging, and the continuum of care. *J. Am. Diet. Assoc.* 96:1048.

10. Position paper of the American Dietetic Association. (1998): Liberalized diets for older adults in long-term care. *J. Am. Diet. Assoc.* 98:201.

11. Prendergast, A. & Fulton, F.L. (1997). *Medical Terminology: A Text/Workbook.* 4th Ed. Redwood City, California. Addison-Wesley Nursing.

12. Pronsky, Z.M. & Solomon, E. (1997). *Food-Medication Interactions.* 10th Ed. Phoenix, Arizona. Food-Medication Interactions, Publishers and Distributors.

13. Saltzman, J.R. & Russell, R.M. (1998). The aging gut. Nutritional issues. *Gastroenterol Clin. North Am.* 27(2):309-324.

14. Schlenker, E.D. (1997). *Nutrition in Aging.* St. Louis, Missouri. Mosby - Year Book, Inc.

15. Whitney, E.N., Cataldo, C.B., & Rolfes, S.R. (1998). *Understanding Normal and Clinical Nutrition,* 5th Ed. West/Wadsworth.

CASE STUDY #13
PEDIATRIC GERD

INTRODUCTION

This case study involves a young neonatal/pediatric patient with GERD that required surgery to correct. The study is designed to introduce the student to some facts about gestation, premature births, and pediatric nutrition.

SKILLS NEEDED

ABBREVIATIONS:

Knowledge of the following abbreviations is required in order to understand this case. You should learn these abbreviations before you begin to read the study.

BUN : blood urea nitrogen
Cl : chloride
Cr : creatinine
GERD : gastroesophageal reflux disease
g/dl : grams per deciliter
Hgb : hemoglobin
IHDP : Infant Health and Development Program
in : inch
K : potassium
lbs : pounds

mEq/L : milliequivalent per liter
mg/dl : milligram per deciliter
ml : milliliter
oz : ounce
PEG : percutaneous endoscopic gastrostomy
po : by mouth
Ser Alb : serum albumin
VLBW : very low birth weight
YOWM : year old white male

LABORATORY VALUES:

You will need to be able to interpret the nutritional significance of the following laboratory values for this case study: BUN, Cr, glucose, Hgb, Na, K, Cl, and Ser Alb.

MEDICATIONS:

Become familiar with the following medications before reading the case study. Note the diet-drug interactions, dosages and methods of administration, gastrointestinal tract reactions, etc.

1. Pepcid (famotidine); 2. Mylanta (simethicone), 3. Propulsid (cisapride), 4. Ventolin (albuterol), 5. Cefzil (cefprozil).

PG is a 2 YOWM who was admitted to the hospital with recurrent problems resulting from GERD. PG was born prematurely at 30 weeks, weighing 2 lbs 7 oz, and was classified as a VLBW baby. He was 15 ins. at birth. Being premature, he was on a ventilator for most of the time he spent in the hospital. After one month, PG was diagnosed with pneumonia. About this time it was noticed that he had difficulty feeding, and although nothing therapeutic was done for this, he was watched closely at feeding time. Over the next several months, he continued to have pulmonary problems and was not growing as he should be expected. This was attributed to his condition and the fact that his intake was not what it should be. PG was discharged after three months in the hospital. At six months, PG was taking solid foods better than liquids but it was noted that he was still having problems swallowing. At six and a half months he was again diagnosed with pneumonia and had to be admitted and treated. He was discharged after treatment with no pulmonary problems. PG's mother did not bring him back for regular check-ups as scheduled, so his progress could not be routinely measured. At two years and nine months she brought him in with breathing problems. Upon examination, he was again diagnosed with pneumonia and was admitted for a series of tests. Chest X-rays indicated aspiration pneumonia. He also underwent a video fluoroscopy which showed him to be at significant risk for aspiration. During the oral phase of the test, PG had functional control over liquids and had fair propulsion with solids, but a delayed trigger swallow was noted during the pharyngeal phase. It was also noted that PG was well behind normal physical and cognitive development. Upon admission, his weight was 16.5 lbs and his height was 28 ins. PG was diagnosed with pneumonia, GERD, and dehydration. His initial nutritionally relevant lab values were as follows:

TEST	RESULT	NORM	TEST	RESULT	NORM
Hgb	13 g/dl	9.5 - 14.1	BUN	19 mg/dl	5 - 18
K	4.6 mEq/L	3.4 - 4.7	Cr	0.4 mg/dl	0.4 - 0.7
Ser Alb	6.3 g/dl	5.9 - 7.0	Na	140 mEq/L	135 - 148
Glucose	95 mg/dl	74 - 127	Cl	99 mEq/L	98 -- 106

After he was treated for pneumonia, PG underwent surgery to have a Nissan fundoplication and a PEG tube placed. PG recovered from surgery without complications and began to receive food via his PEG. Initially, the dietitian recommended that he be fed 3.5 cans of Kindercal (840 ml per day) with Pedialyte as tolerated. This recommendation was not followed but Similac Neocare was used instead, 1180 ml per day. He responded to the feedings without incident. One month after surgery PG was tolerating his feeding well and was showing positive results. He was then offered Similac Neocare po with added rice cereal.

PG's medications included:

Pepcid
Mylanta
Propulsid
Ventolin
Cefzil

****

QUESTIONS:
1.   Answer the following questions with short answers:

   a. What is the normal range of time in weeks for gestation?

   b. Before how many weeks is a baby considered to be premature when born?

   c. What is the average birth weight for babies in the United States?

d. What is considered to be a low birth weight baby?

e. What is considered to be a very low birth weight baby?

2. List the signs that indicate when a baby should be started on solid food.

3. Define the following:

aspiration pneumonia:

video fluoroscopy:

oral phase of swallowing:

delayed trigger swallow:

pharyngeal phase of swallowing:

Pedialyte:

4. What is a Nissan fundoplication and what will it accomplish?

5. Define PEG, describe how it is placed, and give the rationale for its placement in this case.

6. According to the IHDP Growth Percentiles for VLBW Premature Boys (obtainable through Ross Labs; see references 2 and 16), how would you classify PG based on his height and weight for age upon admission for surgery? What should his height and weight be?

7. Knowing that PG is dehydrated, you should be able to predict which lab values would be elevated. List the labs that should be elevated.

8. Cr should be among the labs you listed, but it is not elevated. Give a possible explanation for this in PG's case.

9. What is the caloric and protein requirement for a normal baby that is 2 years and 9 months?

10. What would it be for PG immediately after surgery?

11. After recovering from surgery he will not need extra calories for healing, but he will need extra calories for catch-up growth. What should his caloric and protein intake be at this time?

12. Give the class of each of the following medications and their action.

Medication	Class	Action
Pepcid		
Mylanta		

Propulsid

Ventolin

Cefzil

13.   List any nutritional side effects of the following medications:

Medication   Side Effects
Pepcid

Mylanta

Propulsid

Ventolin

Cefzil

14.   What are the potential consequences of giving children Mylanta for an extended period? [☞ *Hint: Check the list of references*.] Is there cause for concern? Explain.

15.   In the following table, compare Kindercal with Similac Neocare.

Product	Producer	Form	Cal/ml	Non-pro Cal/g N	g/L			Na mg	K mg	mOsm /kg Water	Vol to meet RDAs in ml	g of fiber /L	Free water /L in ml
					Pro	CHO	Fat						
Kindercare													
Similac Neocare													

16.   Compare the kcals and protein PG would receive from 840 cc of Kindercal and 1180 ml of Neocare Similac.

Information Box13 - 1
The original **SOAP** note written in a patient's chart should contain all four components, Subjective, Objective, Assessment, and Plan. However, not every **SOAP** note has to contain all four components. After the original note, subsequent notes could contain from one to four of the components. For example, if additional lab values become available that alter the patient's assessment and plan, but no additional subjective information is available, then a second note could be entered into the patient's chart containing only information in the **OAP** sections. If the patient's nutritional therapy changes but nothing else changes, a third note could be entered into the patient's chart with only the **P** section. See the information box in case study 4, page 37, for additional information.

17.   The first paragraph of page 118 relates that at two years and nine months PG's mother brought him to the doctor with breathing problems. He was admitted for a series of tests. No subjective information is given. Using the objective information presented after that admission and your answers to the previous questions, prepare a SOAP note that includes entries in the objective, assessment, and plan sections.

O:

A:

P:

Related Readings

1.   Akintorin, S.M., Kamat, M., Pildes, R.S., Kling, P., Andes, S., Hill, J., & Pyati, S. (1997). A prospective randomized trial of feeding methods in very low birth weight infants. *Pediatrics.* 100(4):E4.

2.   Casey, P.H., Kraemer, H.C., Bernbaum, J., Yogman, M.W., and Sells, J.C. (1991). Growth status and growth rates of a varied sample of low birth weight, preterm infants: a longitudinal cohort from birth to three years of age. *J. Pediatr.* 119:599-605.

3.   Chohan, N. Senior Editor. (1998). Nursing 98 Drug Handbook. Springhouse Corporation, Springhouse, Pennsylvania.

4.   Cioffi, U., Rosso, L., & De Simone, M. (1998). Gastroesophageal reflux disease. Pathogenesis, symptoms and complications. *Panminerva Med.* 40(2):132-8.

5.   Fischbach, F.T. (1995). *A Manual of Laboratory & Diagnostic Tests.* 5th Ed. Philadelphia. J.B. Lippincott Company.

6.   Foldes, J., Balena, R., Ho, A., Parfitt, A.M., & Kleerekoper. (1991). Hypophosphatemic rickets with hypocalciuria following long-term treatment with aluminum-containing anatacid. *Bone.* 12(2):67-71.

7.   Fomon, S.J. (1993). *Nutrition of Normal Infants.* B.C. Decker.

8.   Hamill, P.V.V., Drizd, T.A., Johnson, C.L., Reed, R.B., Roche, A.F., & Moore, W.M. (1979). Physical growth: National Center for Health Statistics percentilles. *Am. J. Clin. Nutr.* 32:607-29.

9.   *Manual of Clinical Dietetics.* (1996). 5th Ed. Chicago, IL. The American Dietetic Association.

10.   McCance, K., L. & Huether, S.E. (1997). *Pathophysiology: The Biologic Basis for Disease in Adults and Children.* 3rd Ed. Mosby-Year Book.

11.   *Pediatric Nutrition Handbook.* (1998). 4th Ed. American Academy of Pediatrics.

12.   Peters, J.H., DeMeester, T.R., Crookes, P., Oberg, S., de Vos Shoop, M., Hagan, J.A., & Bremner, C.G. (1998). The treatment of gastroesophageal reflux disease with laparoscopic Nissen fundoplication: prospective evaluation of 100 patients with "typical" symptoms. *Ann. Surg.* 228(1):40-40.

13.   Pivnick, E.K., Kerr, N.C., Kaufman, R.A., Jones, D.P., & Chesney, R.W. (1995). Rickets secondary to phosphate depletion. A sequela of antacid use in infancy. *Clin. Pediatr.* 34(2):73-8.

14.   Prendergast, A. & Fulton, F.L. (1997). *Medical Terminology: A Text/Workbook.* 4th Ed. Redwood City, California. Addison-Wesley Nursing.

15.   Pronsky, Z.M. & Solomon, E. (1997). *Food-Medication Interactions.* 10th Ed. Phoenix, Arizona. Food-Medication Interactions, Publishers and Distributors.

16. The Infant Health and Development Program: Enhancing the outcomes of low-birth-weight, premature infants. (1990). *JAMA*. 263(22):3035-3042.

17. Trahms, C.M. & Pipes, P.L. (1997). *Nutrition in Infancy and Childhood*. 6th Ed. McGraw-Hill Company.

18. Queen, P.M., & Lang, C.E. Editors. (1992) *Handbook of Pediatric Nutrition*. Gaithersburg, Maryland. Aspen Publishers.

19. van der Peet, D.L., Klinkenberk-Knol, E.C., Eijsbouts, Q.A., van den Berg, M., de Brauw, L.M., & Cuesta, M.A. (1998). Laparoscopic Nissen fundoplication for the treatment of gastroesophageal reflux disease (GERD). Surgery after extensive conservative treatment. *Surg. Endosc.* 12(9):1159-63.

20. Wetscher, G.J., Glaser, K., Wieschemeyer, T., Gadenstatter, M., Klingler, P., Klinger, A., & Hinder, R.A. (1998). Cispride enhances the effect of partial posterior fundoplication on esophageal peristalsis in GERD patients with poor esophageal contractility. *Dig. Dis. Sci.* 43(9):1986-90.

21. Whitney, E.N., Cataldo, C.B., & Rolfes, S.R. (1998). *Understanding Normal and Clinical Nutrition*, 5th Ed. West/Wadsworth.

CASE STUDY #14
PEDIATRIC BPD

INTRODUCTION

This is the second pediatric case study that is intended to provide an introduction to pediatric terminology, assessment, and nutrition. The physicians in this case were very aggressive in the nutritional treatment of this child.

SKILLS NEEDED

ABBREVIATIONS:

Knowledge of the following abbreviations is required in order to understand this case. You should learn these abbreviations before you begin to read the study.

BPD : broncopulmonary displesia
BM : black male
cm : centimeter
G-tube : same as PEG-tube
IHDP : Infant Health and Development Program
Ht : height
kcal/hr : kilocalories per hour
kg : kilogram
mg : milligram
mg/dl : milligram per deciliter
ml/hr : milliliters per hour

mm^3 : cubic millimeter
mos : months
Na : sodium
oz : ounce
PEG : percutanious endoscopic gastrostomy
PICU : pediatric intensive care unit
po : by mouth
tsp : teaspoon
VLBW : very low birth weight
Wt : weight

MEDICATIONS:

Become familiar with the following medications before reading this case study. Note the diet-drug interactions, dosages and methods of administration, gastrointestinal tract reactions, etc.

1. Aldactone (spironolactone); 2. Propulsid (cisapride); 3. Digoxin (digoxin); 4. Lasix (furosemide); 5. Pepcid (famotidine); 6. Slo-Phyllin (theophylline); 7. Vancocin (vancomycin ydrochloride); 8. Ventolin (albuterol).

LV is a 9 month old BM who was admitted to PICU for ventilator support of BPD. He has had chronic problems with BPD since birth and is oxygen dependent. He has also experienced gastric reflux on occasion and mild hyperkalemia. LV was born premature at 28 weeks gestational age and weighed 1 kg. He was classified as a VLBW baby. He had complications as a result of his prematurity that were primarily respiratory. These included pulmonary emphysema, left tension pneumothorax with chest tube placement, and BPD. He required ventilator support until just prior to discharge and was sent home with oxygen via a nasal cannula.

LV was readmitted at 5 months with additional pulmonary complications. His weight at that time was 3.5 kg and his height was 49.8 cm. His physician was concerned about his weight, so after LV's pulmonary problems cleared up, the physician recommended surgery to have a PEG placed. Other minor surgeries were performed that are common to premature births, i.e. inguinal hernia repair and cryosurgery with laser surgery for premature retinopathy. Similac Neocare was started through the G-tube at a rate of 8.4 Kcal/hr (see reference #4.). LV was discharged at 6 months with a social services consult. The concern was that LV's mother had to work and LV has four siblings at home that also need to be provided for.

At 9 months LV has returned for a check-up and weighs 5.2 kg and is 50.1 cms. The G-tube seems to be helping significantly. He is still having some problems with BPD and currently has another pulmonary infection but his reflux and hyperkalemia are resolving and his prognosis has improved. His current feeding is Similac Neocare at the rate of 15.5 ml/hr with 1 tsp of Polycose powder per oz. LV's current medications include:

Aldactone
Diuril
Slo-Phyillin
Pepcid
Ventolin
Cispride
Lasix
Digoxin
Vancomycin

****

QUESTIONS:

1. Define BPD as to the cause, treatment, and prognosis. What is the importance of nutrition in relation to this disorder?

2. Briefly define:
   Pulmonary emphysema:

   Pneumothorax:

   Cryosurgery:

   Polycose:

3. According to the IHDP Growth Percentiles for VLBW Premature Boys (obtainable through Ross Labs; see references 2 and 17), how would you classify LV for height and weight for age at 5 months? What should his height and weight be at this age?

4.   According to the IHDP Growth Percentiles for VLBW Premature Boys, how would you classify LV for height and weight for age at 9 months? What should his height and weight be at 9 months?

5.   At 5 months of age, LV was receiving Similac Neocare at a rate of 8.4 kcals per hour. How many calories and how much protein was this providing? How many calories and how much protein should he be receiving at 5 months?

6.   At 9 months of age, LV was receiving Similac Neocare at a rate of 15.5 kcals per hour with 1 tsp of Polycose per oz. How many calories and how much protein was this providing? How many calories and protein should he be receiving at 9 months? In your answer show how many mls in an ounce and how much Polycose LV was receiving.

7.   How much free water was LV receiving at 5 months? At 9 months? How much should he be receiving at both 5 and 9 months?

8.   Give the class of each of the following medications and its action.

Medication	Class	Action
Aldactone		
Diuril		
Slo-Phyillin		
Pepcid		
Ventolin		

Medication	Class	Action
Cispride		
Lasix		
Digoxin		
Vancomycin		

9.   List any nutritional side effects of the following medications:

Medication	Side Effects
Aldactone	
Diuril	
Slo-Phyillin	
Pepcid	
Ventolin	
Cispride	
Lasix	
Digoxin	
Vancomycin	

Related Readings

1.   Akintorin, S.M., Kamat, M., Pildes, R.S., Kling, P., Andes, S., Hill, J., & Pyati, S. (1997). A prospective randomized trial of feeding methods in very low birth weight infants. *Pediatrics.* 100(4):E4.

2.   Casey, P.H., Kraemer, H.C., Bernbaum, J., Yogman, M.W., and Sells, J.C. (1991). Growth status and growth rates of a varied sample of low birth weight, preterm infants: a longitudinal cohort from birth to three years of age. *J. Pediatr.* 119:599-605.

3.   Chohan, N., Senior Editor. (1998). Nursing 98 Drug Handbook. Springhouse Corporation, Springhouse, Pennsylvania

4.   Ekvall, S. W., Editor. (1993). *Pediatric Nutrition in Chronic Diseases and Developmental Disorders.* New York. Oxford University Press.

5.   Fischbach, F.T. (1995). *A Manual of Laboratory & Diagnostic Tests.* 5th Ed. Philadelphia. J.B. Lippincott Company.

6.   Fomon, S.J. (1993). *Nutrition of Normal Infants.* B.C. Decker.

7.   Jacob, S.V., Coates, A.L., Lands, L.C., MacNeish, C.F., Riley, S.P., Hornby, L., Outerbridge, E.W., Davis, G.M., & Williams, R.L. (1998). Long-term pulmonary sequelae of severe bronchopulmonary dysplasia. *J. Pediatr.* 133(2):193-200.

9.   *Manual of Clinical Dietetics.* (1996). 5th Ed. Chicago, IL. The American Dietetic Association.

10.   McCance, K.L. & Huether, S.E. (1997). *Pathophysiology: The Biologic Basis for Disease in Adults and Children.* 3rd Ed. Mosby-Year Book.

11.   Mueller, D.H. (1998). Timeliness of codifying ABCDE's for BPD. *J. Pediatr.* 133(3):315-6.

12.   Palta, M., Sadek, M., Barnet, J.H., Evans, M., Weinstein, M.R., McGuinness, G., Peters, M.E., Gabbert, D., Fryback, D., & Farrell, P. (1998). Evaluation of criteria for chronic lung disease in surviving very low birth weight infants. Newborn Lung Project. *J. Pediatr.* 132(1):57-63.

13.   *Pediatric Nutrition Handbook.* (1998). 4th Ed. American Academy of Pediatrics.

14.   Prendergast, A. & Fulton, F.L. (1997). *Medical Terminology: A Text/Workbook.* 4th Ed. Redwood City, California. Addison-Wesley Nursing.

15.   Pronsky, Z.M. & Solomon, E. (1997). *Food-Medication Interactions.* 10th Ed. Phoenix, Arizona. Food-Medication Interactions, Publishers and Distributors.

16.   Singer, L., Yamashits, T., Lilien, L., Collin, M., & Baley, J. (1997). A longitudinal study of developmental outcome of infants with bronchopulmonary dysplasis and very low birth weight. *Pediatrics.* 100(6):987-93.

17. The Infant Health and Development Program: Enhancing the outcomes of low-birth-weight, premature infants. (1990). *JAMA*. 263(22):3035-3042.

18. Trahms, C.M. & Pipes, P.L. (1997). *Nutrition in Infancy and Childhood*. 6th Ed. McGraw-Hill Company.

19. Queen, P.M., & Lang, C.E., Editors. (1992) *Handbook of Pediatric Nutrition*. Gaithersburg, Maryland. Aspen Publishers.

20. Whitney, E.N., Cataldo, C.B., & Rolfes, S.R. (1998). *Understanding Normal and Clinical Nutrition*, 5th Ed. West/Wadsworth.

CASE STUDY #15
PEPTIC ULCER DISEASE

INTRODUCTION
This is a basic study of ulcer disease and involves the treatment of ulcers from two perspectives. Symptoms, treatment, and counseling of the ulcer patient are presented. An emphasis is given to dietary treatment of ulcers, and a right and wrong approach is discussed.

SKILLS NEEDED

ABBREVIATIONS:
Knowledge of the following abbreviations is required in order to understand this case. You should learn these abbreviations before you begin to read the study.

ADHD : attention deficit hyperactive disorder
Alk Phos : alkaline phosphertase
ALT : alanine aminotransferase
ASA : acetylsalicylic acid (aspirin)
AST : aspartate aminotransferase
b.i.d. : twice a day
BRB : bright red blood
BUN : blood urea nitrogen
BUT : biopsy urease test
cc : cubic centimeters
Cl : chloride
cl liqs : clear liquids
CPK : creatine phosphokinase
Cr : creatinine
D_5W : 5% dextrose in water
ELISA : enzyme linked immunosorbent assay
ER : emergency room
g : gram
g/dl : gram per deciliter
h : hour
HCl : hydrochloric acid
Hct : hematocrit

Hgb : hemoglobin
hs : hour of sleep or evening
IgG : immunoglobulin g
I.V. : intravenous
K : potassium
LUQ : left upper quadrant
MCV : mean corpuscular volume
mEq : milliequivalent
mg : milligram
mg/dl : milligram per deciliter
Na : sodium
pc : after meals
po : by mouth
q6h : every 6 hours
RLQ : right lower quadrant
R/O : rule out
Ser Alb : serum albumin
tab : tablet
U/L : units per liter
YOWF : year old white female
x3d : times 3 days
μm^3 : cubic microns

LABORATORY VALUES:
You will need to be able to interpret the nutritional significance of the following laboratory values for this case study: Alk Phos, BUN, Cl, CPK, Cr, glucose, Hct, Hgb, K, MCV, Na, Serum Amylase, Ser Alb, AST, and ALT.

FORMULAS:

The formulas used in this case study include reference body weight, percent reference body weight, the Harris-Benedict equation, and stress factors to determine total caloric needs. The formulas can be found in Appendix A, Tables A - 7 through 10 and Appendix D Tables D - 1 and 5.

MEDICATIONS:

Become familiar with the following medications before reading the case study. Note the diet-drug interactions, dosages and method of administration, gastrointestinal tract reactions, etc.

1. Carafate (sucralfate); 2. AlternaGel (aluminum hydroxide gel); 3. Prilosec (omeprazole); 4. Aspirin (acetylsalicylic acid); 5. TUMS (calcium carbonate); 6. Pepcid (famotidine); 7. Amoxil (amoxicillin); 8. Biaxin (clarithromycin).

GG is a 27 YOWF who was married after one year of college. She has no job skills and has depended on her husband to support her for eight years. Six months ago her husband left her and their 4-year-old son. GG faced the responsibility of supporting herself and her son. She did not receive any financial support from her family and had additional expenses to consider if she were to pursue a divorce settlement, which was necessary to obtain child support for her son. Everyone encouraged her to return to school. Her mother agreed to help by babysitting. GG applied for and received a Guaranteed Student Loan to start school. She obtained a part-time job on campus and returned to school to pursue a degree in accounting.

Adjusting to the new lifestyle was very difficult for GG. Studying was hard for her after such a long time out of school and she had to maintain a good average to keep her loan. The combination of working part-time and trying to be a mother and a father to her son was very demanding. The time demands forced her to eat out frequently and her choices were usually fast foods with a high fat content. Sometimes she did not stop to eat anything. Her home-cooked meals were frequently frozen dinners or fried foods. GG's busy schedule caused her to become fatigued. To overcome this, she started drinking strong, black coffee throughout the day and into the night. GG loved chocolate and usually had chocolate bars during the day for energy. She also started smoking, something she never did before. GG was not sleeping well. To temper this, GG began to have a hot buttered rum (sometimes more than one) at bedtime.

After her husband left, GG started having severe problems with her son that required psychological intervention. He was diagnosed with ADHD and caused GG several problems. GG had always been easily upset and often had to take antacids for a burning stomach. She recently started experiencing an increase in the burning and, in the last month, had severe pain in the RLQ. The pain occurred about 30 minutes after eating. It subsided with antacids but returned. She also had a burning sensation in her RLQ after drinking coffee. Stress headaches were not uncommon. Her remedy for this was ASA daily. Recently, GG felt like she really needed a break and convinced her mother to take care of her son for the weekend to give her some time for herself.

That Friday night GG started drinking hot buttered rum and had a few too many. During the night she awoke with severe stomach pain that radiated to the right side and up into the chest. GG thought she had an indigestion but it was the worst she ever had. She drank some milk because she heard that it would help, and she also took some TUMS. The pain subsided after a while and, still somewhat under the influence of alcohol, she went back to sleep. The next morning she again awoke to severe gastric pain. She tried her earlier remedy and it worked for a while but the pain came back. This continued throughout the morning until she decided to go to the urgent care center and see a doctor. After listening to her symptoms, hardships, and intake of caffeine, alcohol, ASA, cigarettes, and high fat foods, he assumed she had a touch of gastritis that would clear up with medication and dietary changes. He prescribed the following regimen: Cl liqs x3d and then the gradual addition of half and half every 30 minutes; progress to a full liquid diet, then a strict bland diet, continuing with the half and half every 30 minutes. Her medications included: sucralfate (Carafate), 1 tab q6h; aluminum hydroxide gel (AlternaGel), 10 cc po 1h pc and hs; famotidine (Pepcid), 40 mg po hs. He told her to quit drinking alcohol, smoking, and taking aspirin. If the pain did not go away in about a week, she should see her family physician.

****

QUESTIONS:

1. List all the food items that may contribute to GG's condition and explain why.

   fast food with high fat content
   coffee
   alcohol
   fried foods + frozen dinners

2. List any additional oral intake that may have contributed to GG's condition and explain why. *Asprin can cause sudden, serious gastric bleeding, and raises uric acid. The Tums if used excessively can cause abdominal pain*

3. List the non-oral stimulants (physical or psychological stress) that could contribute to GG's condition.

   Her son who was diagnosed with ADHD which caused GG several problems. Studying for school. Keeping up with loans. Major stress

4. List the symptoms of GG's gastritis.

   burning stomach
   abdominal pain
   abdominal pain after drinking

5. Briefly tell what was wrong with GG's diet order and give the scientific basis for your answer. The same principle should apply to the milk and Tums GG was taking.

   Irregular eating patterns. Alcohol and asprin are direct effect on gastritis. Cigarettes have an effect on the acid within stomach. Alcohol Alcohol stimulates acid. Large amount of fat stimulate gastric acid, and high density foods also stimulate acid within the stomach.

6. The Sippy Diet referred to in this case study is very old and hopefully is not found in existing diet manuals. However, there are people who have been told that drinking milk or heavy cream will alleviate symptoms of gastritis or ulcer disease and there will always be those who have been misinformed. If this patient was your responsibility as a dietitian, and you had to deal with this order or an equally inappropriate, scientifically incorrect diet order, discuss how you would approach the physician to correct the order.

7. List the principles of the diet plan that you think GG should follow.

Stop drinking alcohol
Stayaway from fast foods, fried foods, and easy to prepare foods

8. List the physical and psychological changes GG would have to make and give an example of a behavior modification she could use to help her accomplish such a change.

9. What is the mechanism of action of the following medications GG is receiving: Carafate, AlternaGel, and Pepcid?

Carafate acts like a barrier and coats the stomach.
Alternalgel raises the pH in the body
Pepcid histamine blocker, receptor antagonist

10. When should these medications be given in relationship to meals?

Pepcin can be taken any time with no relation to meals, usually at night if once a day. Carafate should be taken before meals or on empty stomach (1 hour before). Alternal gel should be taken 1-3 hours after meal.

11. List the nutrient-drug interactions that are associated with these medications.

Alternagel lowers phosphate, folic acid, and iron. Carafate interferes with calcium and magnesium absorbtion Pepcid interferes with iron absorbtion + vit B12 absorbtion and cannot drink caffeine.

****

GG took the medication as prescribed and refrained from the cigarettes, caffeine, and alcohol as much as possible but, as she started to feel better, went back to her old ways. One night she went on another binge and drank too much. The next morning she started throwing up BRB and not only had RLQ pain but severe LUQ pain as well. This frightened her, so she decided to go to the ER. She was examined and hospitalized to R/O an ulcer.

Her chart indicated the following:

▸ Height 5'2"

▸ Weight 98 lbs

▸ Recent loss of weight: 12 lbs in the last six months.

****

QUESTIONS CONTINUED:

12. What is GG's reference body weight and percent of RBW? RBW = 120

$$\frac{ABW}{RBW} \times 100 = \frac{98}{120} \times 100 = 81.7\%$$

13. Estimate her daily energy needs using the Harris-Benedict equation and appropriate stress factor (Appendix D, Tables D - 1, 2, and 5).

$655.1 + (9.6 \times W) + (1.8 \times ht) - (4.7 \times age)$

$655.1 + (9.6 \times 44.55) + (1.8 \times 13.20) - (4.7 \times 27) =$

$655.1 + 427.68 + 23.78 - 45.9 =$

1060.66

Lab values on admission:

TEST	RESULT	NORM	TEST	RESULT	NORM	TEST	RESULT	NORM
Hgb	20.0 g/dl	12 - 16	AST	32 U/L	5 - 40	MCV	81.0 µm³	82 - 98
Hct	53%	36 - 48	ALT	34 U/L	7 - 56	CPK	106 U/L	96 - 140
Ser Alb	4.5 g/dl	3.9 - 5.0	Na	141 mEq/L	135 - 145	K	5.0 mEq/L	3.5 - 5.3
Glucose	155 mg/dl	65 - 110	Cl	103 mEq/L	98 - 106	Serum Amylase	350 U/L	25 - 125
BUN	35 mg/dl	7 - 18	Cr	1.9 mg/dl	0.6 - 1.3	Alk Phos	160 U/L	17 - 142

Her stool was positive for occult blood and an esophagogastroduodenoscopy revealed gastritis superior to the pyloric sphincter with an ulcer on the dorsal wall of the duodenum, just below the pyloric sphincter. During the endoscopy the physician took a biopsy for *Helicobacter pylori* and carcinoma. Gastric analysis indicated hypersecretion of HCl. An I.V. solution of D_5W was started. Because of the blood in her gut, the physician ordered a backup to the H. pylori biopsy with a blood test to detect serum IgG antibody to H. pylori.

The biopsy for carcinoma was negative but a BUT was positive for H. pylori. The ELISA was also positive. As treatment, the physician ordered omeprazole 20 mg b.i.d; amoxicillin 1 g b.i.d.; and clarithromycin 500 mg b.i.d.

Lab values two days after admission:

TEST	RESULT	NORM	TEST	RESULT	NORM	TEST	RESULT	NORM
Hgb	9.0 g/dl	12 - 16	AST	17 U/L	5 - 40	MCV	78.0 µm³	82 - 98
Hct	30%	37 - 47	ALT	15 U/L	7 - 56	CPK	112 U/L	96 - 140
Ser Alb	2.8 g/dl	3.9 - 5.0	Na	140 mEq/L	135 - 145	K	3.7 mEq/L	3.5 - 5.3
Glucose	90 mg/dl	65 - 110	Cl	105 mEq/L	98 - 106	Serum Amylase	95 U/L	25 - 125
BUN	11 mg/dl	7 - 18	Cr	1.0 mg/dl	0.6 - 1.3	Alk Phos	150 U/L	17 - 142

****

14.   What might be the cause of the LUQ pain along with her usual pain? (☞ *Hint: Consider the enzymes that are elevated.*) *Her Acute pancreatitis from the serum amylase.*

(mild alcoholic hepatitis)

15. Refer to the table of lab values obtained on admission and two days after admission. Note that GG's Hgb, Hct, Glucose, BUN, Cr, Ser Alb, Na, K, and Cl all dropped two days after admission. Explain all the circumstances that may be involved with this.

GG was hydrated during her stay.
Blood loss.

16. Refer to the two lab tables again, and note that two days after admission, GG's Serum Amylase, AST, and ALT all dropped while her Alk Phos and CPK remained essentially unchanged. Explain what this probably indicates.

pancreatitis may be improving, and the liver was healing from the alcholol injury

---

Information Box 15 - 1

It is difficult to culture H. pylori. The usual detection techniques are either the BUT or the serum antibodies test, but not both. There is evidence to indicate that blood in the stomach could produce a false negative[4] for the BUT, so a backup test would not be completely unreasonable. The two tests were ordered here to help the student to become familiar with the tests. The organism is also difficult to eradicate. The combination of medications that is most effective is still under evaluation but the combinations usually include a H_2-antagonist or a proton pump inhibitor, along with a combination of two antibiotics[1,5]. There is a definite theory behind this approach. H. pylori survives in the acidic environment of the stomach by producing urease which deaminates amino acids (hence the reasoning for a BUT). The free amino-group acts as a buffer and raises the pH to 4 to 6. This enables H. pylori to survive, but it is unable to reproduce at this pH. Antibiotics are ineffective if the organism is not replicating, thus, single antibiotic therapy does not eradicate the organism. When a H_2-antagonist or a proton pump inhibitor is included with the antibiotics, the pH is raised to between 6 and 8, which allows the organism to replicate. At this point the antibiotics can destroy the organism[9].

---

17. What diagnostic test(s) (not lab values) indicate(s) that GG has an ulcer?

He did a esophagogastroduodenoscopy and actually visualized the ulcer. The ELISA test which came out positive

18.   Briefly sketch the anatomical position where GG's ulcer can be found.

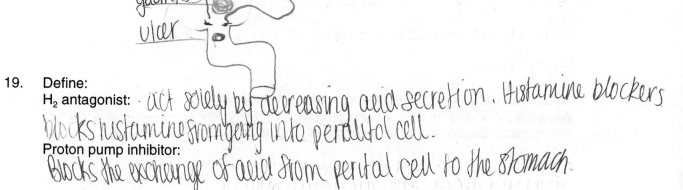

gastritis

ulcer

19.   Define:
H₂ antagonist: act solely by decreasing acid secretion. Histamine blockers blocks histamine from going into peritutol cell.

Proton pump inhibitor:
Blocks the exchange of acid from perital cell to the stomach.

20.   What is the mechanism of action of the following medications GG is receiving: omeprazole, amoxicillin, and clarithromycin? Omeprazole lowers gastric acid secretion and raises gastric pH. amoxicillin kills the bacteria along clarithromycin

Related Readings

1.   Adami, H.O., Bergstrom, R., Nyren, O., Forhaug, K., Gustavsson, S., Loof, L., & Nyberg, A. (1987). Is duodenal ulcer really a psychosomatic disease? A population-based case-control study. *Scand. J. Gastroenterol.* 22(7):889-96.

2.   Anda, R.F., Williamson, D.F., Escobedo, L.G., & Remington, P.L. (1990). Smoking and the risk of peptic ulcer disease among women in the United States. *Arch. Intern. Med.* 150(7):1437-41.

3.   Chu, K.M., Choi, H.K., Tuen, H.H., Law, S.Y., Branicki, F.J., & Wong, J. (1998). A prospective randomized trial comparing the use of omeprazole-based dual and triple therapy for eradication of Helicobacter pylori. *Am. J. Gastroenterol.* 93(9):1436-1442.

4.   Crabtree, J.E., Shallcross, T.M., Heatley, R.V., & Wyatt, J.I. (1991). Evaluation of a commercial ELISA for serodiagnosis of Helicobacter pylori infection. *J. Clin. Pathol.* 44(4):326-8.

5.   Fischbach, F.T. (1995). *A Manual of Laboratory & Diagnostic Tests.* 5th Ed. Philadelphia. J.B. Lippincott Company.

6.   Leung, W.K., Sung, J.J., Siu, K.L., Chan, F.K., Ling, T.K., & Cheng, A.F. (1998). False-negative biopsy urease test in bleeding ulcers caused by the buffering effects of blood. *Am. J. Gastroenterol.* 93(10):1914-8.

7.   Levenstein, S., Kaplan, G.A., & Smith, M. (1995). Sociodemographic characteristics, life stressors, and peptic ulcer. A prospective study. *J. Clin. Gastroenterol.* 21(3):185-92.

8.   Levenstein, S. & Kaplan, G.A. (1998). *J. Clin. Gastroenterol.* 26(1):14-7.

9.   Nursing 98 Books. (1998). *Drug Handbook.* Springhouse, Pennsylvania. Springhouse Corporation.

10.   Pipkin, G.A., Williamson, R., & Wood, J.R. (1998). Review article: one-week clarithromycin triple therapy regimens for eradication of Helicobacter pylori. *Aliment. Pharmacol. Ther.* 12(9):823-37.

11.   Piper, D.W., McIntosh, J.H., Greig, M., & Shy, C.M. (1982). Environmental factors and chronic gastric ulcer. A case control study of the association of smoking, alcohol, and heavy analgestic ingestion with exacerbation of chronic gastric ulcer. *Scand. J. Gastroenterol.* 17(6):721.

12.   Prach, A.T., Malek, M., Tavakoli, M., Hopwood, D., Senior, B.W., & Murray, F.E. (1998). H_2-antagonist maintenance therapy versus Helicobacter pylori eradication in patients with chronic duodenal ulcer disease: a prospective study. *Aliment. Pharmacol. Ther.* 12(9):873-80.

13.   Prendergast, A. & Fulton, F.L. (1997). Medical Terminology: A Text/Workbook. 4th Ed.

14. Pronsky, Z.M. & Solomon, E. (1998). *Food-Medication Interactions.* 10th Ed. Phoenix, Arizona. Food-Medication Interactions, Publishers and Distributors.

15. Scott, D., Weeks, D., Melchers, K., & Sachs, G. (1998). The life and death of Helicobacter pylori. *Gut.* Suppl. 1:S56-60.

16. Svanes, C., Soreide, J.A., Skarstein, A., Fevang, B.T., Bakke, P., Vollset, S.E., Svanes, K., & Sooreide, O. (1997). Smoking and ulcer perforation. *Gut.* 41(2):177-80.

17. Vaira, D., Holton, J., Menegatti, M., Landi, F., Ricci, C., Ali, A., Gatta, L., Farinelli, S., Acciardi, C., Massardi, B., & Miglioli, M. (1998). Blood tests in the management of Helicobacter pylori infection. Italian Helicobacter pylori Study Group. *Gut.* Suppl. 1:S39-46.

18. Whitney, E.N., Cataldo, C.B., and Rolfes, S.R. (1998). *Understanding Normal and Clinical Nutrition,* 5th Ed. West/Wadsworth.

19. Yasunaga, Y., Bonilla-Palacios, J.J., Shinomura, Y., Kanayama, S., Miyazaki, Y., & Matasuzawa, Y. (1997). High prevalence of serum immunoglobulin G antibody to Helicobacter pylori and raised serum gastrin and pepsinogen levels in enlarged fold gastritis. *Can. J. Gastroenterol.* 11(5):433-6.

CASE STUDY #16
GASTRECTOMY

INTRODUCTION

This is a continuation of Case Study #15. GG did not take care of herself and had to have surgery. Additional information about PUD is covered with basic information on surgery. Review information for a gastrectomy and the diet for dumping syndrome before completing this study. Treatment of gastric ulcers is so advanced that surgery is not performed as frequently as it was in the past, but there are still many reasons why someone could be required to have a gastrectomy.

SKILLS NEEDED

ABBREVIATIONS:

Knowledge of the following abbreviations is required in order to understand this case. You should learn these abbreviations before you begin to read the study.

ADHD : attention deficit hyperactive disorder
ASA : acetylsalicylic acid (aspirin)
BUN : blood urea nitrogen
BRB : bright red blood
Cr : creatinine
D_5NS : 5% dextrose in normal saline
ER : emergency room
g/dl : gram per deciliter
Hct : hematocrit
Hgb : hemoglobin
I.V. : intravenous
lap : laparotomy
L : liter
lymph : lymphocytes

MCV : mean corpuscular volume
mg/dl : milligram per deciliter
mm^3 : cubic millimeter
NH_3 : ammonia
P : phosphorous
PC : packed cells
post-op : post operative
PT : prothrombin time
PUD : peptic ulcer disease
sec : seconds
Ser Alb : serum albumin
$\mu mol/L$: micromoles per liter
μm^3 : cubic microns

LABORATORY VALUES:

You will need to be able to interpret the nutritional significance of the following laboratory values for this case study: BUN, Cr, glucose, Hct, Hgb, MCV, PT, NH_3, WBC, lymph, P, and Ser Alb.

FORMULAS:

The formulas used in this case study include calculation of total energy expenditure and protein requirements. The formulas can be found in Appendix D, Tables D - 1, 2, and 5.

MEDICATIONS:

Become familiar with the following medications before reading the case study. Note the diet-drug interactions, dosages and methods of administration, gastrointestinal tract reactions, etc.
1. Prilosec (omeprazole); 2. Amoxil (amoxicillin); 3. Biaxin (clarithromycin).

GG's medications were effective and the bleeding stopped. The RD talked with the physician and could not convince him to change to a soft diet but was able to get him to compromise with a liberal bland diet. GG was discharged and was instructed to continue taking her medications, [Prilosec (omeprazole), Amoxil (amoxicillin), and Biaxin (clarithromycin)], for 14 days with a bland diet. GG followed her diet and took her medications for a while but since she felt so good, she quit taking the medication before she used all of it. She started going to a counselor on campus and her family provided additional help with babysitting. GG made it through the semester. That summer she worked full time and did not try to go to school. She spent more time with her son and that pleased him. The symptoms of ADHD were minimized.

The next fall she started going back to school and took a full load. Her classes were hard and it was not long before GG was again too busy to accomplish everything. Her mother got the flu and was not able to babysit for her. GG's son went to kindergarten in the morning but needed a babysitter in the afternoon. Kindergarten and a babysitter were additional expenses for GG. Her divorce proceedings began and her husband was trying to fight paying the child support her lawyer was asking. GG started back to her old habits. She did not have time to eat right. She was upset most of the time and started smoking again to help calm herself. Staying up late was necessary and so was the coffee. The burning in her stomach started again. GG was determined not to let it slow her down.

She stopped going to her counselor and felt like she had no one to talk to. Since she was so pressed for time, her son was not getting the quality time with her that he had enjoyed during the summer. The symptoms of ADHD once again became intensely evident. GG was receiving pressure from every side. She was desperately trying to cope but could not perform well under pressure. The burning got worse and more frequent. GG was losing weight and started feeling tired and weak. She was not the type of person to quit so she kept trying. GG needed to unwind. Her drinking restarted and so did the ASA. She noted that her stools were getting darker, but she did not see that as important, so she overlooked it.

One evening, after a very stressful day, GG had several drinks and started vomiting BRB. She called her mother who brought her to the ER where she was diagnosed with PUD. An IV was started with D$_5$NS, and GG received several units of PC. An abdominal tap was done and was positive for red blood cells. GG was rushed to surgery with a perforated ulcer.

Her lab values before receiving PC included:

TEST	RESULT	NORM	TEST	RESULT	NORM	TEST	RESULT	NORM
Hgb	8.0 g/dl	12 - 16	NH$_3$	45 µmol/L	9 - 33	MCV	85.0 µm^3	82 - 98
Hct	28%	36 - 48	WBC	12x10^3 /mm^3	5 - 10 x 10^3	Lymph	12.0%	20 - 40
Ser Alb	3.5 g/dl	3.9 - 5.0	BUN	24 mg/dl	7 - 18	Cr	1.0 mg/dl	0.6 - 1.3
PT	11 sec	10 - 13	Control	10.4 sec	--	P	2.3 mg/dl	2.5 - 4.5

Everyone has a breaking point and the events leading up to GG's reaching hers should be obvious. It is important to spot problem areas so that proper counsel can be provided. If you are not

qualified to counsel in a certain area, leave it alone, but be aware of whom you can refer your client to for help.
****

QUESTIONS:

1. GG was not receiving counsel at the time that the major bleeding started. If you had the opportunity to counsel GG just before the bleeding, in what areas would you feel competent to counsel her and in what areas would you refer her to someone else? Investigate the agencies in your area that are available to provide assistance to someone like GG.

2. What is the significance of the dark stools?

Usually means blood that comes from the stomach. The stomach digests the blood and turns it into a black color.

3. Give the pathophysiology for the following abnormal values: BUN, NH_3, and WBC.

BUN is up because of excessive blood in the gut wich is protien and catabolized. WBC values were increased due to inflamation, and stress, Amonia can be a sign of liver failure from the drinking

4. GG was probably dehydrated on admission since she had been drinking. This means that some of her lab values were probably higher/lower (circle one) than indicated.

Higher

5. After admission GG received packed cells and IV fluids. How would that affect the next set of lab values?

Raise The next set of lab values the hemoglobin and hematocrit will be elevated, whereas the BUN will be lowered because of hydration

****

An exploratory lap was done and additional ulceration was found in her stomach just superior to the pyloric sphincter. This was the area that contained gastritis during the previous tests. GG received a partial gastrectomy with a Billroth I and a vagotomy. The ulcer created a small fistula but did not erode a blood vessel. GG was very fortunate; it could have been a lot worse. The post-op period went very well. GG recovered without complications. She was sent home on a postgastrectomy diet and was placed under the care of a psychiatrist.

****

QUESTIONS CONTINUED:

6. Define the following terms:

Packed cells: *are the removal of the blood serum, and consist predominately of the red blood cells*

Abdominal tap: *A needle into the stomach to take fluid out of.*

Perforated ulcer: *A superficial ulcer that goes through the stomach wall*

Fistula: *A development of a tract/tunnel from one organ to another*

Exploratory Laparotomy: *surgery into the abdomen*

Billroth I: *loss of normal regulation of gastric emptying but less dumping than with Billroth II (gastroduodenostomy)*

Vagotomy: *resection or removal of the portion of the vagus nerve innervating the parietal cells for the purpose of diminishing gastric acid secretion*

7. Sketch a Billroth I.

8. Compare a Billroth I to a Billroth II as to anatomical changes as well as to dietary changes, if any.

9. Calculate GG's energy and protein needs.

10. List the principles of a postgastrectomy diet and briefly describe the scientific basis for each principle.

11. Is it possible that GG's diet will ever change or do you believe she will be on a postgastrectomy diet for the rest of her life? Explain your answer.

Related References

1. Chung, C.S. (1997). Surgery and gastrointestinal bleeding. *Gastrointest. Endosc. Clin. N. Am.* 7(4):687-701.

2. Fischbach, F.T. (1995). *A Manual of Laboratory & Diagnostic Tests.* 5th Ed. Philadelphia. J.B. Lippincott Company.

3. Kyzer, S., Binyamini, Y., Melki, Y., Ohana, G., Koren, R., Chaimoff, C., & Wolloch, Y. (1997). Comparative study of the early postoperative course and complications in patients undergoing Billroth I and Billroth II gastrocetomy. *World J. Surg.* 21(7):763-6.

4. Lau, W.Y. & Leow, C.K. (1997). History of perforated duodenal and gastric ulcers. *World J. Surg.* 21(8):890-6.

5. Lee, Y.T., Sung, J.J., Choi, C.L., Chan, F.K., Ng, E.K., Ching, J.Y., Leung, W.K., & Chung, S.C. (1998). Ulcer recurrence after gastric surgery: is Helicobacter pylori the culprit? *Am. J. Gastroenterol.* 93(6):928-31.

6. Leivonen, M., Nordling, S., & Haglund, C. (1998). The course of Helicobacter pylori infection after partial gastrectomy for peptic ulcer disease. *Hepatogastroenterology.* 45(20):587-91.

7. Prendergast, A. & Fulton, F.L. (1997). Medical Terminology: A Text/Workbook. 4th Ed.

8. Ed. Pronsky, Z.M. & Solomon, E. (1998). *Food-Medication Interactions.* 10th Ed. Phoenix, Arizona. Food-Medication Interactions, Publishers and Distributors.

9. Nursing 98 Books. (1998). *Drug Handbook.* Springhouse, Pennsylvania. Springhouse Corporation.

10. Ralphs, D.N. (1981). The dumping syndrome. *Br. J. Clin. Pract.* 35(9):291-3.

11. Vecht, J., Masclee, A.A., & Lamers, C.B. (1997). The dumping syndrome. Current insights into pathophysiology, diagnosis and treatment. *Sacnd. J. Gastroenterol. Suppl.* 223:21-7.

12. Whitney, E.N., Cataldo, C.B., and Rolfes, S.R. (1998). *Understanding Normal and Clinical Nutrition*, 5th Ed. West/Wadsworth.

13. Yamashita, Y., Toge, T., & Adrian, T.E. (1997). Gastrointestinal hormone in dumping syndrome and reflux esophagitis after gastric surgery. *J. Smooth Muscle Res.* 33(2):37-48.

CASE STUDY #19
CHOLECYSTITIS

INTRODUCTION

This is a very basic study of cholecystitis that involves terminology, lab values, and diet. This study also points out that even if a person is overweight, their protein status may not be adequate. Surgery is necessary to remove the gallbladder but is uncomplicated and serves as an introduction for basic surgical care.

SKILLS NEEDED

ABBREVIATIONS:

Knowledge of the following abbreviations is required in order to understand this case. You should learn these abbreviations before you begin to read the study.

Alk Phos : alkaline phosphertase
ALT : alanine phosphatase
AST : aspartate aminotransferase
Bili : bilirubin
BMI : body mass index
BUN : blood urea nitrogen
Cl : chloride
CPK : creatine phosphokinase
Cr : creatinine
g/dl : gram per deciliter
GB : gall bladder
Hct : hematocrit
Hgb : hemoglobin
K : potassium
LDH : lactic dehydrogenase
MCH : mean corpuscular hemoglobin

MCV : mean corpuscular volume
mEq : milliequivalent
mg/dl : milligram per deciliter
mm^3 : millimeter
Na : sodium
NPO : nothing by mouth
P : phosphorous
pg : picogram
RBW : reference body weight
RUQ : right upper quadrant
Ser Alb : serum albumin
TLC : total lymphocyte count
UBW : usual body weight
U/L : units per liter
WBC : white blood cell count
YOWF : year old white female
μm^3 : cubic micrometer

LABORATORY VALUES:

You will need to be able to interpret the nutritional significance of the following laboratory values for this case study: Alk Phos, Bili, BUN, Cl, CPK, Cr, glucose, Hct, Hgb, K, LDH, lymphocytes, MCH, MCV, Na, Ser Alb, Serum Amylase, ALT, AST, TLC, P, and WBC.

FORMULAS:

The formulas used in this case study include reference body weight, percent reference body weight, percent usual body weight, adjusted body weight, body mass index, energy and protein requirements, and total lymphocyte count. The formulas can be found in Appendices A and D, Tables A - 7 through 13, 15, and D - 1 and 5.

Mrs. T is a 45 YOWF who is married with 4 children. She has been healthy most of her life with no history of any major disease. She is 5'6" tall and weighs 195 lbs. Her UBW was 145 lbs most of her adult life. Since her 40th birthday, she has been gaining weight slowly. About a year ago she started having some pain in her RUQ after eating. The pain began to get worse and now her whole abdomen is sore and tender to touch. The abdominal pain is accompanied by a severe pain in the right side of her back, near the upper right shoulder. The pain is so severe at times that it causes Mrs. T to become nauseous. She finally went to her doctor who examined her and took a x-ray. The x-ray showed some opaque stones in the gallbladder. He advised her to go into the hospital for further tests and a cholecystectomy. At first, Mrs. T refused surgery. Not long after that visit, after participating in a large meal with her family to celebrate her 25th wedding anniversary, she began experiencing severe pain in the RUQ, right back shoulder, and was nauseous. She gave in to her husband's urging and agreed to go to the hospital.

A cholecystography and a choledochography were conducted and cholecystitis was diagnosed with a gallstone in the neck of the GB. There was no evidence of stones in the cystic or common bile ducts. Mrs. T was placed on clear liquids and elective surgery was scheduled for Monday, four days later. The next day the diet tech was conducting her routine screening of surgical patients and found the following lab values in the computer for Mrs. T on admission:

TEST	RESULT	NORM	TEST	RESULT	NORM	TEST	RESULT	NORM
Hgb	10.0 g/dl	12 - 16	MCH	24 pg	26 - 34	MCV	80.0 μm^3	82 - 98
Hct	36%	36 - 48	WBC	12.8×10^3 /mm^3	$5 - 10 \times 10^3$	Lympho-cyte	15.0%	20 - 40
Ser Alb	3.3 g/dl	3.9 - 5.0	Na	138 mEq/L	135 - 145	K	3.5 mEq/L	3.5 - 5.3
Glucose	130 mg/dl	65 - 110	Cl	100 mEq/L	98 - 106	P	3.3 mg/dl	2.5 - 4.5
BUN	16 mg/dl	7 - 18	Cr	0.7 mg/dl	0.6 - 1.3	ALT	28 U/L	7 - 56
CPK	120 U/L	96 - 140	LDH	540 U/L	313 - 618	AST	40 U/L	5 - 40
Serum Amylase	120 U/L	25 - 125	Alk Phos	230 U/L	17 - 142	Bili	2.8 mg/dl	0.2 - 1.0

The diet tech turned this information over to the dietitian. The dietitian became concerned after reviewing the screening and decided to visit Mrs. T. The conversation that ensued revealed the following:

Mrs. T's husband travels a lot and is often gone for several days at a time. When Mrs. T reached her 42nd birthday, her last child left home to go to college. As long as Mrs. T had children at home, she was very active with PTA activities, taking the children places, and working around the house. After they all left home, however, Mrs. T found herself approaching middle age and alone most of the time with nothing to do. Mild depression resulted. She found comfort in watching soap operas but could not watch television very long without having something to snack on. She had eaten healthy foods most of her life, but being alone most of the time, she did not have anyone to be a role model for and she began to eat whatever she wanted. Thus, the quality of her snacks deteriorated and consisted of high sugar, high fat foods, all of which had little protein value. As a result, she started gaining weight while her protein status decreased.

As the dietitian talked with Mrs. T, she observed that she appeared to be healthy with indications of protein malnutrition and obesity. She had a medium frame.
****

QUESTIONS:

1.   Determine Mrs. T's RBW and percent of RBW (Appendix A, Table A - 7 through 10).

2.   Using the formula for determining adjusted body weight, calculate her adjusted body weight (Appendix A, Table A - 11).

3.   Calculate Mrs. T's TLC (Appendix A, Table 15).

4.   Calculate Mrs. T's BMI (Appendix A, Tables A - 12 & 13).

5.   List the conditions that probably contributed to Mrs. T's weight gain. Include in your list any possible conditions that may not have been discussed in this case study.

6. Mrs. T is not a small woman. She looks healthy and, other than the cholecystitis symptoms, feels fine. Based on the information given about her diet, how can her obesity be explained with her depressed visceral protein stores?

7. Considering the lab values and the dietitian's observations, would you recommend that Mrs. T have surgery on Monday, after four days of clear liquids? Discuss why or why not.

8. If the surgery was to be postponed, what diet order would you recommend for Mrs. T? Remember to consider her cholecystitis and malnutrition. Would you recommend a mineral supplement? If so, explain which one(s) you would recommend and what dose would be appropriate. Explain your reasoning.

9. Discuss the foods you would recommend to Mrs. T to include in her diet in order to follow your diet order. Remember to be practical!

10.   What energy level would you recommend? Give the rationale for your answer.

11.   After the surgery is completed and Mrs. T has recovered, would you change your diet order for home use? Tell how and explain why.

12.   What practical points could you tell Mrs. T about her lifestyle that may help her keep her weight under control?

13.   Define the following terms:

Cholecystectomy:

Cholecystitis:

Cholecystography:

Choledochography:

Malnutrition:

14.   List the symptoms of cholecystitis.

15. Sketch the location of the GB and identify the hepatic duct, cystic duct, and common bile duct. Identify the location of the stone in the neck of the GB.

16. What are gallstones made of?

17. What relationship, if any, does sex, age, fertility, and weight have on cholecystitis?

18. There are two sets of abnormal lab values here. One set of values is abnormal because of cholecystitis and one because of malnutrition. List all abnormal values and normal values. Indicate if they are abnormal because of cholecystitis or malnutrition.

ABNORMAL VALUE   NORMAL VALUE   CHOLECYSTITIS   MALNUTRITION

19. A modern technique for cholecystectomy is laparoscopic surgery. Compare laparoscopic surgery to conventional surgery in respect to: risk, wound healing, recovery time, diet, and the length of time patient is NPO.

Related References

1. Bordelon, B.M., Hobday, K.A., & Hunter, J.G. (1993). Laser vs. electrosurgery in laparoscopic cholecystectomy. A prospective randomized trial. *Arch. Surg.* 128(2):233-6.

2. Fischbach, F.T. (1995). *A Manual of Laboratory & Diagnostic Tests.* 5th Ed. Philadelphia. J.B. Lippincott Company.

3. Fox, J.G., Dewhirst, F.E., Shen, Z., Feng, Y., Taylor, N.S., Paster, B.J., Ericson, R.L., Lau, C.N., Correa, P., Araya, J.C., & Roa, I. (1998). Hepatic Helicobacter species identified in bile and gallbladder tissue from Chileans with chronic cholecystitis. *Gastroenterology.* 114(4):755-63.

4. Hamy, A., Visset, J., Likholatnikov, D., Lerat, F., Gibaud, H., Savigny, B., & Paineau, J. (1997). Percutaneous cholecystostomy for acute cholecystitis in critically ill patients. *Surgery.* 121(4):398-401.

5. Hawasli, A. (1994). Timing of laparoscopic cholecystectomy in acute cholecystitis. *J. Laparoendosc. Surg.* 4(1):9-16.

6. Prendergast, A. & Fulton, F.L. (1997). *Medical Terminology: A Text/Workbook.* 4th Ed.

7. Pronsky, Z.M. & Solomon, E. (1998). *Food-Medication Interactions.* 10th Ed. Phoenix, Arizona. Food-Medication Interactions, Publishers and Distributors.

8. Ransom, K.J. (1998). Laparoscopic management of acute cholecystitis with subtotal cholecystectomy. *Am. Surg.* 64(10):955-7.

9. Schwesinger, W.H. & Diehl, A.K. (1996). Changing indications for laparoscopic cholecystectomy. Stones without symptoms and symptoms without stones. *Surg. Clin. North Am.* 76(3):493-504.

10. Whitney, E.N., Cataldo, C.B., and Rolfes, S.R. (1998). *Understanding Normal and Clinical Nutrition*, 5th Ed. West/Wadsworth.

CASE STUDY #20
CYSTIC FIBROSIS

INTRODUCTION
This is a study of the genetic disease cystic fibrosis, which usually manifests during infancy or early childhood. The life expectancy for individuals with this disease is about 20 years. The disease involves the mucus-producing glands in the pancreas, bronchi, intestines, and bile ducts. It is more common in whites than blacks, and more common in blacks than Asians.

SKILLS NEEDED

ABBREVIATIONS:
Knowledge of the following abbreviations is required in order to understand this case. You should learn these abbreviations before you begin to read the study.

CF : cystic fibrosis

CI : chloride

mEq/L : milliequivalent per liter

Na : sodium

YOWM : year old white male

LABORATORY VALUES:
You will need to be able to interpret the nutritional significance of the following laboratory values for this case study: Cl, Na, and the iontophoretic sweat test.

MEDICATIONS:
Become familiar with the following medications before reading the case study. Note the diet-drug interactions, dosages and methods of administration, gastrointestinal tract reactions, etc.

1. bronchodilators; 2. corticosteroids; 3. potassium iodine; 4. Viokase (pancrelipase).

RM is a 10 YOWM who has been suffering from CF since he was four years old. His height is 4'3" and he weighs 57 lbs. RM has frequent upper respiratory tract infections and usually has copious secretions of mucous. His parents take excellent care of him, but in spite of their attempts to keep RM healthy, he has recurring respiratory problems. When he was four years old, the disease manifested itself with some abdominal pain, distention, steatorrhea, and a persistent cough that produced copious amounts of sputum. At that time, an iontophoretic sweat test was conducted and the results were:

▸ Na 90 mEq/L

▸ Cl 60 mEq/L.

As RM grew older the symptoms became severe, with respiratory tract infections being the predominant manifestation. RM's medications include bronchodilators, corticosteroids, and potassium iodide daily. When RM has a respiratory attack, he also receives the appropriate antibiotics and respiratory treatments (daily). His physicians are hoping to prevent bronchiectasis, atelectasis, and pneumonia, fatal complications for most CF patients. To help the digestion of his

food, RM is receiving pancrelipase (Viokase). His digestion has not been as much of a problem as the respiratory attacks. Steatorrhea is no longer a problem with RM as long as he takes his enzymes. He has a ravenous appetite and does very well with his diet. When RM has respiratory distress, he gets very anxious, and this affects his digestion and requires an increase in the dosage of pancrelipase.
****

QUESTIONS:

1.   What should RM's height and weight be for his age?

2.   What percentile is he in now for height and weight?

3.   Define the following terms:

    Steatorrhea:

    Bronchiectasis:

    Atelectasis:

    Pneumonia:

4.   Discuss the genetic possibility of being born with CF based on gender and race.

5.   Explain the effects of CF on the body.

6.   What are the goals for nutritional therapy for CF?

7. What information about RM would be helpful in order to determine the percentages of energy that should come from carbohydrates, fats, and proteins?

8. What is a source of fat that would be helpful in the diet of a person with CF?

9. What is the mechanism of action of the following drugs?

   pancrelipase (Viokase):

   potassium iodide:

   corticosteroid:

10. List any nutritional complications or interactions with other drugs.

11. What is the purpose of giving bronchodilators and corticosteroids?

12. Is there any probability that these drugs would have an effect on the diet? Explain your answer.

13.   Explain the iontophoretic sweat test.

► ADDITIONAL OPTIONAL QUESTIONS ◄

Tube Feeding Drill:

14.   If RM were to become seriously ill, require hospitalization, and need to be fed via a feeding tube, what characteristics would be appropriate for the tube feeding you would use?

15.   Using the table below, compare several of the enteral nutritional supplements that would be appropriate for a seriously ill CF patient (Appendix F).

Product	Producer	Form	Cal/ ml	Non- pro Cal/g N	g/L			Na mg	K mg	mOsm /kg Water	Vol to meet RDAs in ml	g of fiber /L	Free water /L in ml
					Pro	CHO	Fat						

Related References

1. Anthony, H., Bines, J., Phelan, P., & Paxon, S. (1998). Relation between dietary intake and nutritional status in cystic fibrosis. *Arch. Dis. Child.* 78(5):443-7.

2. Benabdeslam, H., Garcia, I., Bellon, G., Gilly, R., & Revol, A. (1998). Biochemical assessment of the nutritional status of cystic fibrosis patients treated with pancreatic enzyme extracts. *Am. J. Clin. Nutr.* 67(5):912-8.

3. Fischbach, F.T. (1995). *A Manual of Laboratory & Diagnostic Tests.* 5th Ed. Philadelphia. J.B. Lippincott Company.

4. Jelalian, E., Stark, L.J., Reynolds, L., & Seifer, R. (1998). Nutrition intervention for weight gain in cystic fibrosis: a meta analysis. *J. Pediatr.* 132(3):486-92.

5. Quirk, P.C., Ward, L.C., Thomas, B.J., Holt, T.I., Shepherd, R.W., & Cornish, B.H. (1997). Evaluation of bioelectrical impedance for prospective nutritional assessment in cystic fibrosis. *Nutrition.* 13(5):412-6.

6. Murphy, J.L. & Wootton, S.A. (1998). Nutritional managemant in cystic fibrosis--an alternative perspective in gastrointestinal function. *Disabil. Rehabil.* 20(6-7):226-34.

7. Navarro, J., Munck, A., & Varille, V. (1995). Energy balance and nutritional support in cystic fibrosis. *Pediatr. Pulmonol. Suppl.* 11:74-5.

8. Nursing 98 Books. (1998). *Drug Handbook.* Springhouse, Pennsylvania. Springhouse Corporation.

9. Prendergast, A. & Fulton, F.L. (1997). *Medical Terminology: A Text/Workbook.* 4th Ed.

10. Pronsky, Z.M. & Solomon, E. (1997). *Food-Medication Interactions.* 10th Ed. Phoenix, Arizona. Food-Medication Interactions, Publishers and Distributors.

11. Rendina, E.A., Venuta, F., DeGiacomo, T., Guarino, E., Ciccone, A.M., Quattrucci, S., Rocca, G.D., Antonelli, M., Ricci, C., & Coloni, G.F. (1998). Lung transplantation for cystic fibrosis. *Eur. J. Pediatr. Surg.* 8(4):208-11.

12. Stallings, V.A., Fung, E.B., Hofley, P.M., & Scanlin, T.F. (1998). Acute pulmonary exacerbation is not associated with increased energy expenditure in children with cystic fibrosis. *J. Pediatr.* 132(3):493-9.

13. Wallace, C.S., Hall, M., & Kuhn, R.J. (1993). Pharmacologic management of cystic fibrosis. *Clin. Pharm.* 12(9):657-74; quiz 700-1.

14. Whitney, E.N., Cataldo, C.B., and Rolfes, S.R. (1998). *Understanding Normal and Clinical Nutrition,* 5th Ed. West/Wadsworth.

15. Wilson, D.C. & Pencharz, P.B. (1998). Nutrition and cystic fibrosis. *Nutrition.* 14(10):792-5.

CASE STUDY #23
MYOCARDIAL INFARCTION

INTRODUCTION
This study concerns the nutritional implications of someone who has a myocardial infarction but does not have hypercholesterolemia. The nutritional modifications involve a sodium restricted diet. Another ethnic group is introduced.

SKILLS NEEDED

ABBREVIATIONS:
Knowledge of the following abbreviations is required in order to understand this case. You should learn these abbreviations before you begin to read the study.

ALP : alkaline phosphatase
ALT : alanine aminotransferase
AST : aspartate aminotransferase
BMI : body mass index
BUN : blood urea nitrogen
Ca : calcium
CCU : coronary care unit
Cl : chloride
C/O : complains of
CPK : creatinine phosphokinase
CPK_{1-3} : isoenzymes of creatine phosphokinase
Cr : creatinine
Dx : diagnosis
EKG : electrocardiogram contractions
g : gram
g/dl : grams per deciliter
Hct : hematocrit
Hgb : hemoglobin
IU : international units
K : potassium

LD_{1-5} : isoenzymes of lactic dehydrogenase
LDH : lactic dehydrogenase
MCV : mean corpuscular volume
mEq/L : milliequivalents
Mg : magnesium
mg/dl : milligram per deciliter
MI : myocardial infarction
mm^3 : cubic millimeter
MS : morphine sulfate
Na : sodium
NPO : nothing by mouth
O_2 : oxygen
prn : as needed
PVCs : premature ventricular
RBW : reference body weight
U/L : units per liter
WBC : white blood cell count
YO : year old
μm^3 : cubic microns

FORMULAS:
The formulas used in this case study include total calorie and protein needs using appropriate stress factor (Appendix D, Tables D - 1, 2, and 5).

LABORATORY VALUES:
You will need to be able to interpret the nutritional significance of the following laboratory values for this case study: BUN, Cl, Ca, Mg, CPK, CPK_1 (CPK-BB), CPK_2 (CPK-MB), CPK_3 (CPK-MM), Cr, Glucose, Hct, Hgb, K, LDH, LD_1, LD_2, LD_3, LD_4, LD_5, MCV, Na, ALT, AST, ALP, and WBC (Appendix B).

MEDICATIONS:

Become familiar with the following medications before reading the case study. Note the diet-drug interactions, dosages and methods of administration, gastrointestinal tract reactions, etc.
1. Lactated Ringer's; 2. morphine sulfate; 3. Norpace (disopyramide phosphate); 4. Sectral (acebutolol); 5. nitroglycerin; 6. Barbita (phenobarbital).

Mr. Y is a 52 YO Japanese-American. He came to the United States as a college student 34 years ago. He is married and has two children. Mr. Y is a very successful computer programer for a major firm. He has worked his way up to a high level of management and is now in a very stressful position. He is very good at what he does but is in a highly competitive market. His division must produce in order for the company to survive and he feels that it is largely his responsibility to see that this is accomplished. Because great emphasis was placed on excellence in his upbringing, he has always been an over-achiever. He must not only do well, he must be the best. He is married to a woman from Japan. Though he has been in this country for 34 years, he still follows many of the practices of his homeland. His diet is greatly Americanized but many aspects of the Japanese diet are still evident. In his work he has the opportunity to fly to Japan on occasion, so he maintains strong ties with his homeland. He is 5'5" and weighs 155 lbs. There is no family history of cardiovascular disease but he has a family history of intestinal cancer. He has been treated for ulcers twice.

It is very important to him that he continues to be a success in his work. He has become a model for his family in America and in Japan. This places additional stress on him and he stays at the office late at night and brings work home with him in the evenings and on weekends. One day at work he had a heart attack. He began to feel very weak, turned pale, and felt like he had pressure pushing in on his chest. It was difficult to breathe. There was severe pain in the sternum area that radiated to his left shoulder and down his arm, almost to the elbow. He also had pain radiate up his neck to his jaw. He collapsed to the floor and was rushed to the hospital. He was admitted to the CCU with a Dx of MI. When Mr. Y was admitted, he still felt a crushing sensation on his chest with some burning and pain in the sternum area. He was diaphoretic, pale, was having PVCs, and was very anxious. His temperature was 99.9°. His initial treatment included O_2, MS, and lactated Ringer's. He was NPO. The EKG suggested that Mr. Y had had an MI and that he was in ventricular tachycardia. Mr. Y was given disopyramide phosphate (Norpace), acebutolol (Sectral), and nitroglycerin prn. Phenobarbital was used to keep Mr. Y calm. Blood was drawn eight hours after admission and analyzed. The results were as follows:

TABLE 1

TEST	RESULT	NORM	TEST	RESULT	NORM	TEST	RESULT	NORM
Hgb	15.0 g/dl	14 - 17.4	MCV	92 µm^3	82 - 98	Ca	9.2 mg/dl	8.6 - 10.0
Hct	45%	42 - 52	WBC	16.4x10^3 /mm^3	5 - 10 x 10^3	Na	140 mEq/L	135 - 145
K	4.2 mEq/L	3.5 - 5.3	Cl	100 mEq/L	98 - 106	AST	54 U/L	5 - 40
BUN	15 mg/dl	7 - 18	Cr	1.0 mg/dl	0.6 - 1.3	Mg	1.4 mEq/L	1.3 - 2.1
Glucose	155 mg/dl	65 - 110	CPK	193 U/L	38 - 174	ALT	50 U/L	7 - 56

LDH and CPK Isoenzymes:

TABLE 2

TEST	RESULT	NORM	TEST	RESULT	NORM	TEST	RESULT	NORM
Total LDH	420 U/L	313 - 618	LD_1	26 %	17 - 27	LD_2	30 %	29 - 39
LD_3	20 %	19 - 27	LD_4	14 %	8 - 16	LD_5	8.3 %	6 - 16
CPK_1	0 IU/L	0	CPK_2	15 IU/L	0 - 7	CPK_3	95 IU/L	96 - 100

After 25 hours blood was drawn again and the 24 hr labs were as follows:

TABLE 3

TEST	RESULT	NORM	TEST	RESULT	NORM	TEST	RESULT	NORM
Total LDH	735 U/L	313 - 618	LD_1	50 %	17 - 27	LD_2	37 %	29 - 39
LD_3	21 %	19 - 27	LD_4	13 %	8 - 16	LD_5	8.5 %	6 - 16
CPK_1	0 IU/L	0	CPK_2	35 IU/L	0 - 7	CPK_3	77 IU/L	96 - 100
ALT	60 U/L	7 - 56	AST	180 U/L	5 - 40	ALP	54 U/L	17 - 142

****

QUESTIONS:

1. List the lab values for the labs drawn at 8 hrs (**TABLES 1 and 2**) that indicate that Mr. Y had had an MI.

    CPK elevated
    CPK2 elevated
    LDH elevated

2. List the lab values for the labs drawn at 24 hrs (**TABLE 3**) that indicate that Mr. Y had had a MI.

    LDH elevated
    CPK2 elevated

3.   Explain the variance between the two sets of labs. In your explanation, tell how the various enzymes are used to distinguish between MI, liver disease, pulmonary disease, etc.

↑ CPK is an early indicator of a MI.

↑ LDA may be a later indicator of MI

CPK 2 is a more specific value for the heart mussle

4.   Give the mechanisms of action of the following drugs:

Norpace (disopyramide phosphate): antiarythmic

nitroglycerin: coronary atery dialator

Sectral (acebutolol): beta blocker treduces adrenalin it ♡ slows it down.

Barbita (phenobarbital): sedative

morphine sulfate: pain relief

5.   What nutritional complications could occur with the listed medications?

Norpace (disopyramide phosphate): Anorexia

nitroglycerin: hypotension

Sectral (acebutolol):

Barbita (phenobarbital): raise in metabolism of vit D + vit K

morphine sulfate: Anorexia

6.   List any important adverse reactions that could occur with these drugs:

Norpace (disopyramide phosphate): blurred vision headach dizziness

nitroglycerin: headache dizziness hypotension

Sectral (acebutolol):

Barbita (phenobarbital): mild respiratory depression, hyperactivity

morphine sulfate: respiratory depression, hypotension drowsiness, sedation, dizziness & weakness

7.   List the symptoms of an MI.

8.   Define the following terms:

Lactated Ringer's:

Ventricular Tachycardia: *bottom ♡ beating rapidy (ventrical)*

Diaphoretic: *sweaty*

****

Mr. Y started to feel better by the next day, but he had occasional chest pain that required MS and nitroglycerin. This continued for the first two days of his hospitalization. By the third day, he was feeling much better. Clear liquids, with no hot or cold liquids, were ordered but he did not feel like eating. He said that his stomach felt bloated and the liquids just stayed there. He was still receiving nitroglycerin for chest pain but much less frequently. He continually C/O a distended abdomen. By the fourth day, he tried to drink more of the clear liquids. He gradually advanced to a full liquid and then a 2 g Na, low-fat diet with no hot or cold beverages on his tray.
****

QUESTIONS CONTINUED:

9.   What does a distended abdomen mean? List the possible causes of this.

10.   What medications may be having an effect on Mr. Y's distended bloated feeling?

11.   Why would Mr. Y not be allowed to have any hot or cold liquids with his meals?

****

Mr. Y continued to improve and was moved out of CCU to a cardiac floor. During his rehabilitation in the hospital the dietitian started teaching him his diet. When the dietitian interviewed him, she found that Mr. Y ate the following types of foods:

His intake of dairy products such as, milk, cheese, and ice cream was satisfactory. Among the protein foods, he ate a variety of fish and shellfish, beef, pork, and chicken. He ate tofu occasionally. As part of his Japanese culture, he also ate whole dried fish, including the bones, and raw fish on occasion. His diet included large quantities of rice and rice products. Among the rice products were mochiko (a flour used to make rice cakes). He and his wife also frequently ate other breads and crackers, somen, millet, and barley. A large variety of vegetables, including eggplant, cucumbers, mushrooms, bamboo shoots, and even some types of seaweed, were part of his diet. These products were cooked with soy sauce, salt, horseradish, dried celery, parsley, and onions. He did not mention fruit or bread products. Mr. Y was advised to follow a Step I diet with a 2 g Na restriction.

****

QUESTIONS CONTINUED:

12.   If you were the dietitian who was going to counsel Mr. Y, list the foods mentioned above that would need to be eliminated from his diet.

13.   Give examples of foods that he might substitute for those that are not suggested on his 2 g Na, low fat diet.

14.   Determine Mr. Y's RBW and percent of RBW (Appendix A, Tables A - 7 through 10).

15. Calculate Mr. Y's BMI (Appendix A, Tables A - 12 and 13).

$$\frac{wt\ kg}{ht\ m^2} =$$

16. Calculate Mr. Y's energy and protein needs post MI and recommend a caloric intake level for him. Show what stress factor you would use (Appendix D, Tables D - 1, 2 and 5).

66 + (13.7 × wt) + (5 × ht) - (6.8 × age) =   30% energy

17. Define the following foods and list the main nutrients they contribute to the diet:

tofu:

somen:

mochiko:

seaweed:

bamboo shoots:

millet:

mushrooms:

18. The kinds of food eaten by Mr. Y have been listed but not the amounts. Based on the information given, evaluate Mr. Y's diet.

19. Is the 2 g Na diet prescription too strict? Discuss the reasoning behind your answer.

20.   Describe a Step I diet.

▸ **ADDITIONAL OPTIONAL QUESTIONS**◂

21.   Prepare a SOAP note for Mr. Y.

22.   Considering all of the dietary modifications (Step I diet with a 2 g Na restriction) Mr. Y should be practicing and his known dietary habits, plan a day's menu.

Related References

1. Albert, C.M., Hennekens, C.H., O'Donnell, C.J., Ajani, U.A., Carey, V.J., Willett, W.C., Ruskin, J.N., & Manson, J.E. (1998). Fish consumption and risk of sudden cardiac death. *JAMA*. 279(1):23-8.

2. Fischbach, F.T. (1995). *A Manual of Laboratory & Diagnostic Tests*. 5th Ed. Philadelphia. J.B. Lippincott Company.

3. Kim, K.K., Yu, E.S., Liu, W.T., Kim, J., & Kohrs, M.B. (1993). Nutritional status of Chinese-, Korean-, and Japanese-American elderly. *J. Am. Diet. Assoc.* 93(12):1416-22.

4. Kittler, P.G. & Sucher, K.P. (1998). *Food and Culture in America*. 2nd Ed. Belmont, CA. West/Wadsworth.

5. Lappalainen, R., Kiokkalainen, M., Julkunen, J., Saarinen, T., & Mykkanen, H. (1998). Association of sociodemographic factors with barriers reported by patients receiving nutrition counseling as part of cardiac rehabilitation. *J.Am. Diet. Assoc.* 98(9):1026-9.

6. Morton, N.E., Gulbrandsen, C.L., Rao, D.C., Rhoads, G.G., & Kagan, A. (1980). Determinants of blood pressure in Japanese-American Families. *Hum. Genet.* 53(2):261-6.

7. Nursing 98 Books. (1998). *Drug Handbook*. Springhouse, Pennsylvania. Springhouse Corporation.

8. Prendergast, A. & Fulton, F.L. (1997). *Medical Terminology: A Text/Workbook*. 4th Ed.

9. Pronsky, Z.M. & Solomon, E. (1998). *Food-Medication Interactions*. 10th Ed. Phoenix, Arizona. Food-Medication Interactions, Publishers and Distributors.

10. Whitney, E.N., Cataldo, C.B., and Rolfes, S.R. (1998). *Understanding Normal and Clinical Nutrition*, 5th Ed. Belmont, CA. West/Wadsworth.

CASE STUDY #24
CONGESTIVE HEART FAILURE

INTRODUCTION
This is a continuation of case study #23 and follows up on Mr. Y's MI. Through no fault of his own, Mr. Y's condition worsened and resulted in congestive heart failure. He is very sick and is not able to eat. A feeding tube has been placed to maintain nutritional status. Review notes on congestive heart failure and basics about concentrated tube feedings.

SKILLS NEEDED

ABBREVIATIONS:
Knowledge of the following abbreviations is required in order to understand this case. You should learn these abbreviations before you begin to read the study.

BEE : basal energy expenditure	MI : myocardial infarction
BMI : body mass index	MS : morphine sulfate
BS : bowel sounds	N/G : nasogastric
cc : cubic centimeter	NPO : nothing by mouth
CCU : coronary care unit	NTG : nitroglycerin
CHF : congestived heart failure	prn : as needed
C/O : complains of	qd : every day
D/C : discontinue	q4h : every 4 hours
Dx : diagnosis	RBW : reference body weight
I.V. : intravenous	TF : tube feeding
K : potassium	tid : three times a day

FORMULAS:
The formulas used in this case study include total calorie and protein needs using the Harris-Benedict equation and appropriate stress factor, and BMI. (Appendices A, Tables A - 7, 8, 12, 13 and Appendix D, Tables D - 1, 2, and 5).

MEDICATIONS:
Become familiar with the following medications before reading the case study. Note the diet-drug interactions, dosages and methods of administration, gastrointestinal tract reactions, etc.

1. Dobutrex (dobutamine hydrochlorine); 2. morphine sulfate; 3. Norpace (disopyramide phosphate); 4. Lasix (furosemide); 5. nitroglycerin; 6. Barbita (phenobarbital); 7. Slow-K (potassium chloride).

Mr. Y continued to do well over the next several days, but then had another episode of severe chest pain. He required MS and NTG. He was nauseous, pale, cold, and very anxious. He had some chest pain but it was not as severe as it was in his first attack. He was readmitted to CCU. The chest pain recurred frequently and required several doses of MS. He was made NPO and C/O a distended abdomen. The tests conducted indicated that Mr. Y had had another MI. His renal output decreased greatly and Mr. Y was diagnosed with CHF and had to be intubated. His new orders included the following:

► 1. D/C disopyramide phosphate (Norpace)
► 2. NPO
► 3. MS q4h prn
► 4. phenobarbital
► 5. dobutamine hydrochloride (Dobutrex)
► 6. NTG
► 7. furosemide (Lasix)
► 8. 1000 cc total fluid qd including I.V.

****

QUESTIONS:

1. Describe the mechanism of action of the following drugs:

   Dobutrex (dobutamine hydrochloride):

   Lasix (furosemide).

2. Are there any nutritional complications that could occur with these drugs? If so, describe the complications. *yes furosemide causes anorexia and raises your thurst so inturn could cause dehydration.*

3. What does intubation mean?

****

Mr. Y improved over the next few days but was still having problems with CHF. He could not talk because of the endotracheal tube but still indicated that he felt distended, even though a N/G tube was in place to lower suction. Mr. Y has been in the hospital for 14 days and had solid food for 5 of those days. He has been NPO for the last 4 days. Mr. Y has very faint BS and an abdominal radiographic study suggests that he has a gastric ileus. The physician decided to place a feeding tube into the small bowel and start a TF.

****

Information Box 24-1

When a patient requires a feeding tube, in most cases, the optimum placement of the feeding tube is into the small bowel. This is done for the following reasons: First, if an ileus were to occur in a patient after surgery or trauma (including an MI), the ileus would usually be gastric or colonic. The small intestines are usually still functioning during this time. Therefore, a feeding entering the small bowel could be effective without as much of a chance of aspiration. Second, if something were to happen to allow too much tube feeding to flow into the patient, there is less of a chance of aspiration if the tube is in the small bowel.

The strength and flow rate of the feeding depends on a number of factors: the osmolality of the feeding, the position of the tip of the feeding tube, the length of time since the patient has eaten, the endemotus state of the patient's gut, and any disease state affecting digestion or absorption. The more concentrated the feeding and the longer it has been since the patient has eaten, the slower the rate of administration and the greater the dilution of the feeding.

QUESTIONS CONTINUED:

4. Explain the reasoning behind the placement of the feeding tube into the small intestines in Mr. Y's case.

5. Calculate Mr. Y's RBW (Appendix A, Tables A - 7 and 8).

6. Calculate Mr. Y's BEE using the Harris-Benedict equation. Calculate his total daily energy expenditure using the appropriate stress factor (Appendix D, Tables D - 1 , 2, and 5).

7. What are the goals of nutritional therapy for someone with CHF?

8.   Did any of Mr. Y's requirements change with the new Dx? If so, list the changes and explain the factors that caused the changes.

9.   What would be an appropriate kind of TF to start for Mr. Y? Remember that Mr. Y is in CHF and is on restricted fluids and remember where the tip of the feeding tube is.

10.   At what rate and strength would you start Mr. Y's TF? Explain.

11.   What final rate and strength would you try to achieve? Show the progression that you would use each day from your initial rate and strength to your final rate and strength.

12.   Define an ileus.

13.   Speculate on the possible cause of Mr. Y's gastric ileus. Can you give a suggestion that could help alleviate this ileus?

****

Because of the Lasix, Mr. Y's urinary output has increased considerably, and he is gradually getting over CHF. However, the Lasix has also caused his K to drop. Since K given by I.V. is painful, could cause phlebitis, and is required to be given slowly with lots of fluid, the physician decided to give a K supplement via feeding tube. He ordered the patient to have Slow-K (potassium chloride) t.i.d. The next day, Mr. Y's K was in the normal range and the physician continued the Slow-K. By the next day, Mr. Y started to experience some diarrhea and his serum K dropped again. The physician increased the Slow-K by feeding tube. The diarrhea increased and the serum K dropped further.
****

QUESTIONS CONTINUED:

14.   Explain the physiological results of giving Slow-K via Mr. Y's feeding tube.

15.   Define the following terms:

        Endotracheal tube:

        Radiographic:

        Phlebitis:

        N/G to low suction:

► **ADDITIONAL OPTIONAL QUESTIONS** ◄

Tube Feeding Drill:

16. Using the table below, compare several of the enteral nutritional supplements that would be appropriate for Mr. Y's condition (☞ *Hint: remember that fluids and sodium are a problem and that the tip of the feeding tube is in the small bowel.*)

Product	Producer	Form	Cal/ml	Non-pro Cal/g N	g/L			Na mg	K mg	mOsm /kg Water	Vol to meet RDAs in ml	g of fiber /L	Free water /L in ml
					Pro	CHO	Fat						

Related References

1.   Fischbach, F.T. (1995). *A Manual of Laboratory & Diagnostic Tests.* 5th Ed. Philadelphia. J.B. Lippincott Company.

2.   Kirsten, R., Nelson, K., Kirsten, D., Heintz, B. (1998). Clinical pharmacokinetics of vasodilators. Part II. *Clin. Pharmacokinet.* 35(1):9-36.

3.   Kittler, P.G. & Sucher, K.P. (1998). *Food and Culture in America.* 2nd Ed. Belmont, CA. West/Wadsworth.

4.   Nursing 98 Books. (1998). *Drug Handbook.* Springhouse, Pennsylvania. Springhouse Corporation.

5.   Prendergast, A. & Fulton, F.L. (1997). *Medical Terminology: A Text/Workbook.* 4th Ed.

6.   Pronsky, Z.M. & Solomon, E. (1997). *Food-Medication Interactions.* 10th Ed. Phoenix, Arizona. Food-Medication Interactions, Publishers and Distributors.

7.   Skipper, A. (1998). *Dietitian's Handbook of Enteral and Parenteral Nutrition.* 2nd Ed. Gaithersburg, Maryland. Aspen Publishers.

8.   Vesely, D.L., Dietz, J.R., Parks, J.R., Baig, M., McCormick, M.T., Clinton, G., & Schocken, D.D. (1998). Vessel dilator enhances sodium and water excretion and has beneficial hemodynamic effects in persons with congestive heart failure. *Circulation.* 98(4):323-9.

9.   Whitney, E.N., Cataldo, C.B., and Rolfes, S.R. (1998). *Understanding Normal and Clinical Nutrition,* 5th Ed. West/Wadsworth.

CASE STUDY #25
HYPOGLYCEMIA

INTRODUCTION

This study is designed to help the student understand the various reasons why someone may experience hypoglycemia. This disease could be an early indication of diabetes, but most people who experience hypoglycemia usually do so as a result of poor nutrition. Many think they have hypoglycemia but actually have nothing more than poor eating habits.

SKILLS NEEDED

ABBREVIATIONS:

Knowledge of the following abbreviations is required in order to understand this case. You should learn these abbreviations before you begin to read the study.

CHO : carbohydrate
FBS : fasting blood sugar
g : gram
GTT : glucose tolerance test
h : hour

mg/dl : milligram per deciliter
min : minute
YOBF : year old black female
2° : secondary

LABORATORY VALUES:

You will need to be able to interpret the nutritional significance of the following laboratory values for this case study: FBS and GTT (Appendix B).

Mrs. J is a 30 YOBF who is married and has four children, ages 5, 6, 8, and 10. Her husband is unemployed and has been for several months. The only work that Mrs. J has been able to obtain is cleaning or cooking in someone's home. Mrs. J works two or three days a week while her husband takes care of the children. They have to depend on food stamps, city subsistence programs, and their family to meet their needs. They are trying hard to make it but times are difficult and they cannot do any better right now. They are embarrassed about their situation and do not like having to depend on anyone else to meet their needs.

Mrs. J has been feeling dizzy and weak. Recently, while grocery shopping, she got so weak that she had to sit down. Someone gave her a soda and in a little while she felt better. Her husband convinced her that she should see a doctor, in spite of the cost. He was afraid that she had high blood pressure since it runs in her family. Her father died of a stroke and she has a sister with renal failure due to high blood pressure.

She went to the doctor who examined her and found her blood pressure to be normal. However, her blood sugar was very low, 48 mg/dl. To Mrs. J's knowledge, there is no family history of diabetes or hypoglycemia. She remembers her parents saying that her grandmother had some kind of "blood disease," but she does not know what that meant. The doctor's examination revealed the following: Mrs. J has not been eating as she usually does because of the low cash flow. She has been eating a lot of high carbohydrate foods that are inexpensive and easy to fix, such as rice, spaghetti, etc. She has also been eating a lot of bread with her meals because it is filling. About two hours after eating she becomes dizzy, weak, and anxious, sweats profusely, and just "does not feel well." If she lies down the symptoms usually pass. Mrs. J also expressed that she is very worried about her family's situation and usually gets upset after one of their "cheap meals" because she believes she is not adequately providing for her family. The MD thought the problem was reactive, functional hypoglycemia based on a high carbohydrate intake accompanied by hyperepinephrinemia due to nervousness. To be sure, he ordered a 5 hour GTT. To prepare for the test, Mrs. J had to be on a diet of at least 150 grams of CHO and had to report early in the morning without eating after midnight. Mrs. J had a hard time drinking the sweet solution (75 g glucose) on an empty stomach. Having blood drawn every hour on an empty and nervous stomach made Mrs. J sick. She started throwing up after about three hours and the test had to be stopped. The results at that time were:

GTT

Time	0 min	30 min	1 hr	2 hr	3 hr
Glucose mg/dl	70	130	160	90	54

The physician felt that this was sufficient to make a diagnose of reactive functional hypoglycemia. He gave her a sedative and referred her to the dietitian. He assured her that her problem could be resolved by diet and by dealing with her emotions.

The dietitian talked with Mrs. J and determined that she was on a high-carbohydrate, moderate-fat, low-protein diet. The carbohydrate was a mixture of complex carbohydrates and a significant concentration of simple sugars. For the amount of total carbohydrate she was ingesting, the fiber content was low. She consumed three meals per day with frequent high carbohydrate snacks. The snacks were typically caffeine-containing sodas, candy, cookies, and coffee with sugar. Her total intake of caffeine was high.

The dietitian counseled her and questioned her to see if she understood the diet. When her responses were satisfactory, she sent her home with printed material and a number to call if she had questions.
****

QUESTIONS:

1. Define the following terms: hypoglycemia and hyperepinephrinemia.

low levels of blood glucose that lead to neuroglycopenia symptoms which are ameliorated by ingestion of carbohydrates

2. Describe the effects of hyperepinephrinemia.

3.   Mrs. J had a 5 hour GTT. Explain why some GTTs are for 2 h, 3 h, or 5 h.

4.   List the symptoms of hypoglycemia.

blood glucose levels become low
sweating hunger
shaking headaches
weakness irritability

5.   Explain the differences between reactive functional hypoglycemia, reactive hypoglycemia secondary to diabetes, and hypoglycemia secondary to fasting.

6.   What are the normal values for a GTT?

Fasting 70-105   2 hour ≤ 140
5min 300-400   ≥ 3 hour 70-140
30min 180-200
1hour 160-180

7.   Explain the results of Mrs. J's GTT in relationship to her family background.

8.   What is the difference between GTT and gtt?

9.   From the information obtained by the dietitian, list the practices and foods in Mrs. J's lifestyle that could contribute to hypoglycemia.

High carbohydrate foods

10.   List the principles of a diet to prevent hypoglycemia.

Almost no research has been done to determine what food type is related to the treatment and prevention of hypoglycemia.

11.   What specific points would you try to make to Mrs. J for the prevention of hypoglycemia?

12.   The case study indicates three things the dietitian did at the end of her interview that are important. List these three things and explain their importance.

Related References

1. Chandler, P.T. (1977). An update on reactive hypoglycemia. *Am. Fam. Physician.* 16(5):113-6.

2. Comi, R.J. (1993). Approach to acute hypoglycemia. *Endocrinol. Metab. Clin. North Am.* 22(2):247-62.

3. Fischbach, F.T. (1995). *A Manual of Laboratory & Diagnostic Tests.* 5th Ed. Philadelphia. J.B. Lippincott Company.

4. Hofeldt, F.D. (1989). Reactive hypoglycemia. *Endocrinol. Metab. Clin. North Am.* 18(1):185-201.

5. Nursing 98 Books. (1998). *Drug Handbook.* Springhouse, Pennsylvania. Springhouse Corporation.

6. Prendergast, A. & Fulton, F.L. (1997). *Medical Terminology: A Text/Workbook.* 4th Ed.

7. Pronsky, Z.M. & Solomon, E. (1997). *Food-Medication Interactions.* 10th Ed. Phoenix, Arizona. Food-Medication Interactions, Publishers and Distributors.

8. Whitney, E.N., Cataldo, C.B., and Rolfes, S.R. (1998). *Understanding Normal and Clinical Nutrition*, 5th Ed. West/Wadsworth.

CASE STUDY #26
TYPE I DIABETES MELLITUS

INTRODUCTION

This is a basic study concerned with insulin-dependent diabetes mellitus. Study the symptoms, treatment (including the types of insulin used for diabetes), and your diabetic exchanges before you complete this case. A knowledge of carbohydrate counting would also be helpful. Try to understand what is going through this young man's mind. He has never been sick before and has a strong desire to be a professional basketball player. He fears that this is no longer going to be possible.

SKILLS NEEDED

ABBREVIATIONS:

Knowledge of the following abbreviations is required in order to understand this case. You should learn these abbreviations before you begin to read the study.

CDE : certified diabetes educator
DKA : diabetes ketoacidosis
D_5W : 5% dextrose in water
IDDM : insulin-dependent diabetes mellitus
I.V. : intravenous
JODM : juvenile onset diabetes mellitus
MDI : multiple daily injections

mg/dl : milligrams per deciliter
NPH : neutral protamine hagedron insulin
qAM : every morning
R insulin : regular insulin
2° : secondary
YOWM : year old white male

LABORATORY VALUES:

Look up the normal value for glucose and study its relationship to diabetes.

FORMULAS:

The formulas used in this case study include weight for age and total caloric needs.

MEDICATIONS:

Become familiar with the following medications before reading the case study. Note the diet-drug interactions, dosages and methods of administration, gastrointestinal tract reactions, etc.

1. insulin: bovine; porcine; Humulin; NPH, lispro (humulog), and Regular.

JJ is a 13 YOWM who has high hopes of being a basketball star. He is tall for his age and is very well coordinated. He loves the game and plays every chance he gets. He has been healthy until recently. He was playing basketball after school with some friends when he became weak and nauseous. His friends noticed that he was sluggish and not himself, but they thought he was having a bad day. He bent over and rested his hands on his knees and then collapsed. When his friends rushed to him, he was disoriented and confused. He had been playing hard but not that hard. They tried to lay him down and give him some water, but he was getting worse. He was perspiring excessively. They called his parents who then called an ambulance.

JJ was rushed to the hospital and was quickly diagnosed with DKA 2° to Type I diabetes (formerly called Juvenile Onset Diabetes Mellitus [JODM] or Insulin Dependent Diabetes Mellitus [IDDM]). His blood sugar was 620 mg/dl. They started an I.V. with D_5W, R insulin, potassium, and phosphorus. Soon JJ was awake and back to his normal self. It was necessary for him to remain in the hospital for a while to regulate his insulin, diet, and activity.
****

QUESTIONS:

1. Explain the physiology behind the following symptoms:

   Weak, sluggish:

   Nauseous:

   Disoriented and confused:

2. Describe the pathophysiology of DKA.

Diabetic ketoacidosis characterized by elevated blood glucose level and presence of ketones in the urine and the blood. The body depends on fat for energy and ketones are formed.

3. List all of the possible symptoms of Type I DM. Affected persons usually are lean have abrupt onset of symptoms before the age of 30, and are dependent on exogenous insulin to prevent ketoacidosis and death.

4. Explain the use of R Insulin, glucose, potassium, and phosphorous in the I.V. treatment.

****

When the physician talked with JJ's mother, he found out that JJ has a family history of diabetes. His grandfather had diabetes, and he has an uncle with diabetes. His mother further explained that JJ had not been as active as usual for the past several days, but he was eating more frequently. She said JJ complained of losing a couple of pounds in two days but she did not believe his weighing was accurate. He was weighing himself daily because he was trying to gain weight to play ball. He is 5'8" and weighs 140 lbs. She did not notice anything else that was unusual. JJ said he had not felt right recently. He tired easily and was always hungry. He thought it was because he had been playing so hard. JJ's blood sugar was down the second day after his admission, and his serum glucose was 180 mg/dl. He is currently on 5 units of Humulin R insulin and 20 units of Humulin N insulin qAM. He is to have three meals per day with an AM and hs snack. His energy intake level and snacks for exercise are to be determined by the RD, who is also a CDE. The diabetes teaching nurse is to visit him and instruct him in insulin administration and foot care.

****

QUESTIONS CONTINUED:

5. How does JJ's height and weight compare with the norm for his age?

106 + 5(8) = 146 height and weight fit all into the norm

6. What is/are the difference(s) between lispro, R Insulin and NPH insulin?

lispro short acting insulin (clear) onset 5-15 peak 30-75 duration 23
R-insulin short-acting (clear) onset 30-45 peak 2-3 duration 4-6
NPH Background insulin (cloudy) onset 2-4 peak 4-10 duration 10-18

7. What are the differences between Humulin, porcine, and bovine insulin?

8. What is the significance of the AM and hs snacks?

to eat at consistant times which will synchronize the insulin

9. Why is it important for the RN to teach him foot care?

vascular problems which involves circulation

****

JJ and his mother are now very concerned about his future. JJ has heard horror stories about

diabetes and is worried that his basketball career has ended before it started. He knows that his grandfather had to have a leg amputated because of diabetes. Many questions are going through his head when the dietitian comes into his room to talk to him about his diet. His fears are mounting and he has not had the opportunity to talk about this with any professional. Assume that you were that dietitian: how would you handle these questions from JJ?
****

QUESTIONS CONTINUED:

10.   Will I have to take insulin for the rest of my life?

11.   Will I have to go on a special diet and eat diet food?

12.   Will I still be able to eat out with my friends, like at the Bigger Burger; and what would I eat at birthday parties and stuff?

13.   Why did my grandfather lose a leg? Could that happen to me?

14.   Can I still play basketball in high school and college?

15.   Will I be able to gain weight and get a lot stronger?

****
JJ will soon be officially starting basketball practice for his junior high school team. This means that

for the next several months he will be playing hard almost every day and, assuming he makes the team, will be playing in games once or twice a week. He likes fast food and drinks a lot of sodas. When he goes home from school, he likes to snack on cookies, peanut butter and crackers, etc. He is not a picky eater and will eat anything. The dietitian explained the Exchange Lists for Meal Planning to JJ and his mother and encouraged him to start learning the exchanges, particularly the carbohydrate exchanges. She set up an appointment for him in the Out-Patient Diabetic Clinic and told him that the dietitian there will teach him how to do "carbohydrate counting" instead of the "exchanges." She told him that the carbohydrate counting is more flexible. The use of lispro for intensive insulin control through multiple daily injections (MDI) was explained as a possibility also.
****

QUESTIONS CONTINUED:

16. What energy level would you recommend for JJ? Explain how you arrived at your decision. 5'8, 140, 13

66 + (13.7 #wt) + (5*ht) + (6.8*age)

17. What would you teach JJ about the kind and amount of snacks he can have?

When your using a rapid insulin therapy will allow you for flexable times in meals and snacks, as well as the amount of food eaten.

18. What behavioral changes would you recommend to JJ?

Information Box 26-1

In 1993 the published results of the Diabetes Control and Complications Trial (DCCT) brought about several changes in the nutritional treatment and insulin management of diabetes (there are numerous references to this trial listed at the end of this chapter). As a result of the DCCT, there are now two prominent methods of teaching diabetics how to eat to control their blood glucose levels. The older method, the *Exchange Lists for Meal Planing,* is still being used by some who refuse to change. In practice, it is sometimes more suitable for the obese diabetic because it helps them keep track of total calories from sources other than carbohydrate exchanges. The newer method, *Carbohydrate Counting*, is more flexible and provides a method for calculating insulin dosages.

In counseling diabetics, the first step is to determine which system is best for that particular patient. Next, estimate the appropriate energy level. From this, the diet is arranged in exchanges based on a detailed recall or food record.

When counting carbohydrate to balance insulin, the exchange lists are used as a guide but not with the same detail. Example: According to the exchange lists, a bread and fruit exchange are equal to 15 grams of carbohydrate each. A milk exchange is equal to 12 grams of carbohydrate. In the *Counting Carbohydrate* method, the milk exchange is rounded up to 15 grams and all three exchanges, bread, fruit, and milk, are each equal to one *Carbohydrate Counting* exchange of 15 grams. Any food item that contains less than 5 grams of carbohydrate per serving is free if only one serving is eaten at a meal. If more than one serving is eaten, the total amount of carbohydrate in all the servings are added up and compared to the *Carbohydrate Counting* exchange. Example: One vegetable exchange (1/2 cup) at a meal contains 5 grams of carbohydrate and is considered free. Three vegetable exchanges (1 1/2 cups) at a meal contains 15 grams of carbohydrate and is equal to one *Carbohydrate Counting* exchange.[1]

Using the *Carbohydrate Counting* method, there are at least two means to determine the ratio of carbohydrate intake to insulin administered. These are the Carbohydrate Gram Method and the Carbohydrate Choices Method[1]. To teach these methods properly would take more space than allowed here. The references indicated below are a good places to look for further instruction[1].

[1] A very basic explanation of the Carbohydrate Counting Method can be found in a set of three booklets available through the American Dietetic Association or the American Diabetes Association. The booklets, Carbohydrate Counting, Level 1, Level 2, and Level 3 are intended for patients and professionals alike. See references 22 - 24.

Assume the energy level you estimated for JJ in question 16 was exactly correct and assume you decided to use the Carbohydrate Counting method to teach JJ his diet. Based on the answer in question 16 and the information given about JJ's food preferences, i.e., hamburgers, answer question 19.

19.   Plan a balanced diet for JJ for one day. Include snacks at the appropriate time in accordance with his exercise routine. Make the meal plan realistic for a 13-year-old (a good outside assignment would be to interview an athletic 13-year-old and use a "real" diet plan).

20.   From this meal plan determine the carbohydrate exchanges for each meal and snack.

21.   Assume JJ has decided to use intensive insulin control with multiple daily injections (MDI) and he is using 8 units of R insulin in the AM, 4 at lunch, and 6 at the evening meal. Using the two methods for determining a carbohydrate/insulin ratio, determine JJ's ratio.

22.   The current recommendations allow for the incorporation of sugar into a diabetic diet with caution not to indulge to excess. The assumption is that 15 grams of carbohydrate is 15 grams of carbohydrate. Based on JJ's questions, it would be easy to "scare" him into following a strict, or closely controlled diet. How would you counsel JJ specifically about sugar in his diet without using "scare" tactics?

▸ ADDITIONAL OPTIONAL QUESTION ◂

23.   Create a SOAP note for JJ.

Related References

1. Ahmed, A.B. & Home, P.D. (1998). Optimal provision of daytime NPH insulin in patients using the insulin analog lispro. *Diabetes Care.* 21(10):1707-13.

2. American Diabetes Association. (1993). Implications of the Diabetes Control and Complications Trial. American Diabetes Association. *Diabetes.* 42(11):1555-8.

3. American Diabetes Association. (1994). Nutrition recommendations and principles for people with diabetes mellitus. *J. Am. Diet. Assoc.* 94:504.

4. Anderson, J.W. & Geil, P.B. (1998). New perspectives in nutrition management of diabetes mellitus. *Am. J. Med.* 85(5A):159-65.

5. Davis, D.L. & Gregory, R.P. (1993). Carbohydrate counting alternative in glucose control. *J. Am. Diet. Assoc.* 93(10):1104.

6. Del Sindaco, P., Ciofetta, M., Lalli, C., Perriello, G., Pamanelli, S., Torlone, E., Brunetti, P. & Bolli, G.B. (1998). Use of the short-acting insulin analogue lispro in intensive treatment of type 1 diabetes mellitus: importance of appropriate replacement of basal insulin and time-interval injection-meal. *Diabet. Med.* 15(7):592-600.

7. Diabetes Care. (1995). Implication of treatment protocols in the Diabetes Control and Complications Trial. *Diabetes Care.* 18(3):361-76.

8. Dillinger, Y. & Yass, C. (1995). Carbohydrate counting in the management of diabetes. *Diabetes Edu.* 21(6):547-50, 552.

9. Eastman, R.C., Siebert, C.W., Harris, M., & Gorden, P. (1993). Clinical review 51: Implications of the Diabetes Control and Complications Trial. *J. Clin. Endocrinol. Metab.* 77(5):1105-7.

10. Faro, B. (1995). Students with diabetes: implications of the Diabetes Control and Complications Trial for the school setting. *J. Sch. Nurs.* 11(1):16-21.

11. Fischbach, F.T. (1995). *A Manual of Laboratory & Diagnostic Tests.* 5th Ed. Philadelphia. J.B. Lippincott Company.

12. Gregory, R.P. & Davis, D.L. (1994). Use of carbohydrate counting for meal planning in type I diabetes. *Diabetes Educ.* 20(5):406-9.

13. Gillespie, S.J., Kulkarni, K.D., & Daly, A.E. (1998). Using carbohydrate counting in diabetes clinical practice. *J. Am. Diet. Assoc.* 98(8):897-905.

14. Hadden, D.R. (1994). The Diabetes Control and Complications Trial (DCCT): what every endocrinologist needs to know. *Clin. Endocrinol.* 40(3):293-4.

15. Hentzen, D.H. (1994). From the president: the results of the Diabetes Control and Complications Trial. *Diabetes Educ.* 20(2):103.

16.   Koivisto, V.A. (1998). The human insulin analogue insulin lispro. *Ann. Med.* 30(3):260-6.

17.   Nursing 98 Books. (1998). *Drug Handbook.* Springhouse, Pennsylvania. Springhouse Corporation.

18.   Pastors, J.G. (1992). Alternatives to the exchange system for teaching meal planning to persons with diabetes. *Diabetes Educ.* 18(1):57-63.

19.   Prendergast, A. & Fulton, F.L. (1997). *Medical Terminology: A Text/Workbook.* 4th Ed.

20.   Pronsky, Z.M. & Solomon, E. (1998). *Food-Medication Interactions.* 10th Ed. Phoenix, Arizona. Food-Medication Interactions, Publishers and Distributors.

21.   Ronnemaa, T. & Viikari, J. (1998). Reducing snacks when switching from conventional soluble to lispro insulin treatment: effects on glycaemic control and hypoglycaemia. *Diabet. Med.* 15(7):601-7.

22.   The American Dietetic Association. (1995). *Level 1. Carbohydrate Counting: Getting Started.* Chicago, IL.

23.   The American Dietetic Association. (1995). *Level 2. Carbohydrate Counting: Moving On.* Chicago, IL.

24.   The American Dietetic Association. (1995). *Level 3. Carbohydrate Counting: Using Carbohydrate/Insulin Ratios.* Chicago, IL.

25.   The Diabetes Control and Complications Trial Research Group. (1993). The effect of intensive treatment of diabetes on the development and progression of long-term complications in insulin-dependent diabetes mellitus. The Diabetes Control and Complications Trial Research Group. *N. Engl. J. Med.* 329(14):977-86.

26.   Whitney, E.N., Cataldo, C.B., and Rolfes, S.R. (1998). *Understanding Normal and Clinical Nutrition*, 5th Ed. West/Wadsworth.

16. Mahoney, M.C. & Michalek, A.M. (1998). Health status of American Indians/Alaska Natives: genera; patterns of mortality. *Fam. Med.* 30(3):190-5.

17. Nursing 98 Books. (1998). *Drug Handbook.* Springhouse, Pennsylvania. Springhouse Corporation.

18. Peterson, K.P., Pavlovich, J.G., Goldstrin, D., Little, R., England, J., & Peterson, C.M. (1998). What is hemoglobin A1c? An analysis of glycated hemoglobins by electrospray ionization mass spectrometry. Clin. Chem. 44(9):1951-8.

19. Prendergast, A. & Fulton, F.L. (1997). *Medical Terminology: A Text/Workbook.* 4th Ed.

20. Pronsky, Z.M. & Solomon, E. (1998). *Food-Medication Interactions.* 10th Ed. Phoenix, Arizona. Food-Medication Interactions, Publishers and Distributors.

21. Roubicek, M., Vines, G., & Gonzalez, S.A. (1998). Use of HbA1c in screening for diabetes. Diabetes Care. 21(9):1577-9.

22. Whitney, E.N., Cataldo, C.B., and Rolfes, S.R. (1998). *Understanding Normal and Clinical Nutrition*, 5th Ed. West/Wadsworth.

23. Wiener, K. & Roberts, N.B. (1998). The relative merits of haemoglobin A1c and fasting plasma glucose as first-line diagnostic tests for diabetes mellitus in nonpregnant subjects. *Diabet. Med.* 15(7):558-63.

CASE STUDY #28
ALTERNATIVE TREATMENTS for TYPE II DIABETES

INTRODUCTION

This is a continuation of Case Study #27. Mrs. R's treatment was not effective and her condition worsened. Several new treatments are attempted before one is found that works. The new treatments involve the use of the latest drugs for Type II diabetes.

SKILLS NEEDED

ABBREVIATIONS:

Knowledge of the following abbreviations is required in order to understand this case. You should learn these abbreviations before you begin to read the study.

Bili : bilirubin
BS : blood sugar
CDE : certified diabetes educator
Chol : cholesterol
D/C'd : discontinued
g/dl : grams per deciliter

HbA1c : glycosylated hemoglobin
HDL : high density lipoprotein
LDL : low density lipoprotein
mg/dl : milligrams per deciliter
R : regular insulin
Trig : triglycerides
u : units

LABORATORY VALUES:

You will need to be able to interpret the nutritional significance of the following laboratory values for this case study: Glucose, cholesterol, LDL, HDL, triglycerides, HbA1c, ALT, AST, and bilirubin (Appendix B).

FORMULAS:

The only formula used in this case study is the calculation of LDL from the lipid profile. (Appendix D, Table D - 6).

MEDICATIONS:

Become familiar with the following medications before reading the case study. Note the diet-drug interactions, dosages and methods of administration, gastrointestinal tract reactions, etc.

1. insulin: Humulin and Regular; 2. Glucotrol (glipizide); 3. Precose (acarbose); 4. Rezulin (troglitazone); 5. Glucophage (metformin).

Mrs. R took her medication and stayed on her diet as best she could but many of her symptoms continued. She returned to her doctor for regular checkups, the first being one month later. Her weight had increased by ten pounds to 200. Her labs were as follows:

Table 1

TEST	RESULT	NORM	TEST	RESULT	NORM	TEST	RESULT	NORM
Chol	275 mg/dl	140 - 199	Trig	300 mg/dl	40 - 160	HDL	28 mg/dl	40 - 85
Glucose	280 mg/dl	65 - 110	HbA1c	12%	5.5 - 8.5	LDL	187 mg/dl	<130 mg/dl

There was not a significant change in her HbA1c but her weight had increased. Her glucose showed slight improvement. The lipid profile was better in some respects but worse in others. Her original dose of Glucotrol was 5 mg in the morning 30 minutes before breakfast. The physician thought she needed some time to adjust to the new regimen, so he increased her dose to 10 mg and had her report back in two weeks. In the meantime, she was coming to the clinic weekly to see the CDE dietitian and nurse and was instructed to show them her daily log of BS levels from her finger sticks. Over the next several months, Mrs. R continued to report to her doctor with little success. Her weight continued to increase with slight increases in her glucose and HbA1c. Her lipid profile continued to be worrisome. The physician continued to increase her Glucotrol until she was at two doses of 15 mg each. On her last visit her labs were as follows:

Table 2

TEST	RESULT	NORM	TEST	RESULT	NORM	TEST	RESULT	NORM
Chol	262 mg/dl	140 - 199	Trig	323 mg/dl	40 - 160	HDL	26 mg/dl	40 - 85
Glucose	284 mg/dl	65 - 110	HbA1c	12.2%	5.5 - 8.5	LDL	?	<130 mg/dl

Her weight had increased to 205 pounds. With this lack of improvement, the physician decided to start her on insulin, 15 u of Humulin N every morning before breakfast with a sliding scale for regular insulin if her morning glucose was more than 200.

Information Box 28-1
The sliding scale formula the physician used was as follows: if the BS was more than 200, subtract 200 from the actual BS value and divide by two. The final number would be the amount of R insulin used. Example: If the BS was 230: 230 -200 = 30   30/2 =15   Mrs. R. would add 15 u of R insulin to her morning dose.

****

QUESTIONS:

1.   Determine Mrs. R's LDL level in Table 2 (Appendix D, Table D - 6).

2.   Define Humalin N and R insulin.

3.   After injection, when do Humalin N and R insulin begin to have an effect, how long do the effects last, and when do they peak?

Insulin	Onset	Peak	Duration
Humulin N			
R			

****

Mrs. R had to learn how to give herself insulin and did not like it at all. The physician told her that if she got her glucose under control and lost weight, she would not need the insulin. This motivated her to follow her diet and her walking routine. However, as time passed, her condition continued to worsen. She gained weight and her glucose was still elevated. Her HbA1c was still elevated above 12%. Her physician added Precose to her regimen. This immediately had an effect on Mrs. R's BS but too much of an effect. It caused her BS to drop below 60 and gave Mrs. R signs of hypoglycemia. She also complained about excessive gas, abdominal cramps and diarrhea. The physician increased her Humulin N to 40 u every morning, left the sulfonylurea the same, and D/C'd the Precose.

Time continued to pass and Mrs. R continued to gain weight. Her BS was going up instead of going down and her Hb1c was increasing. The physician decided she was having insulin resistence and D/C'd the sulfonylurea and replaced it with Rezulin, 400 mg every morning. Before starting her on Rezulin, the physician did some more blood work. The resultswere as follows:

Table 3

TEST	RESULT	NORM	TEST	RESULT	NORM	TEST	RESULT	NORM
Chol	245 mg/dl	140 - 199	Trig	298 mg/dl	40 - 160	HDL	28 mg/dl	40 - 85
Glucose	270 mg/dl	65 - 110	HbA1c	12.7%	5.5 - 8.5	LDL	157 mg/dl	<130 mg/dl
ALT	40 U/L	7 - 56	AST	32 U/L	5 - 40	Bili	0.5 mg/dl	0.2 - 1.0

He was still concerned about her lipid profile and her HbA1c. He advised Mrs. R to continue her diet and told her that she would need to have the above lab work done monthly for the next six months.

****

QUESTIONS CONTINUED:

4.   What is the action of Precose and what are its side effects?

5.   Explain the sudden drop in BS after starting the Precose.

6.   Define sulfonylurea:

7.   Describe fully the process of insulin resistence.

8.   List the aspects of Mrs. R's case that indicate insulin resistence.

9.   What is the action of Rezulin and what are its side effects?

10.   Why did the physician choose to follow the lab values in Table 3?

11.   List the normal levels of HbA1c for nondiabetics and the levels of good, moderate, and poor control for diabetics.

**Mr

Mrs. R finally had favorable results with her new regimen but only after her Rezulin had to be increased to 600 mg per day. The BS levels were coming down along with her HbA1c. Triglycerides came down also but total cholesterol went up slightly with an increase in LDL. Her HDL went up but not proportionately with the LDL. Her last blood test indicated a rise in ALT. The physician was most concerned about her lipid profile so he decided to make another change to a drug that is supposed to help lower triglycerides, total cholesterol, and LDL while raising HDL. He therefore D/C'd the Rezulin and started Glucophage. This treatment was the most successful and helped to get Mrs. R's BS, lipids, and HbA1c under control. Her weight loss was slight but would continue if she followed her diet and developed an exercise program.

****

12.   What is the action of Glucophage and what are its side effects?

13.   This has been a long ordeal for Mrs. R with many opportunities for her to become discouraged. Try to imagine what it would be like to be her dietitian and to be counseling her through this ordeal. What advice would you give Mrs. R?

Related References

1. American Diabetes Association. (1993). Implications of the Diabetes Control and Complications Trial. American Diabetes Association. *Diabetes.* 42(11):1555-8.

2. American Diabetes Association. (1994). Nutrition recommendations and principles for people with diabetes mellitus. *J. Am. Diet. Assoc.* 94:504.

3. Anderson, J.W. & Geil, P.B. (1998). New perspectives in nutrition management of diabetes mellitus. *Am. J. Med.* 85(5A):159-65.

4. Bloomgarden, Z.T. (1998). International Diabetes Federation meeting, 1997. Issues in the treatment of type 2 diabetes; sulfonylureas, metformin, and troglitazone. *Diabetes Care.* 21(6):1024-6.

5. Brown, S.L., Pope, J.F., Hunt, A.E., & Tolman, N.M. (1998). Motivational strategies used by dietitians to counsel individuals with diabetes. *Diabetes Educ.* 24(3):313-8.

6. Buse, J.B., Gumbiner, B., Mathias, N.P., Nelson, D.M., Faja, B.W., & Whitcomb, R.W. (1998). Troglitazone use in insulin-treated type 2 diabetic patients. The Troglitazone Insulin Study Group. *Diabetes Care.* 21(9):1455-61.

7. Cherryman, G., Campbell, S., Crozier, A., Daintith, H., Jeyapalan, K., Keal, R., Lister, D., Reek, C., Upton, D., & Hudson, N. (1998). Diabetics on Metformin. *Clin. Radiol.* 53(6):465.

8. Chiasson, J.L., Gomis, R., Hanefeld, M., Josse, R.G., Karasik, A., and Laakso, M. (1998). The STOP-NIDDM Trail: an international study on the efficacy of an alpha-glucosidase inhibitor to prevent type 2 diabetes in a population with impaired glucose tolerance: rationale, design, and preliminary screening data. Study to Prevent Non-Insulin-Dependent Diabetes Mellitus. *Diabetes Care.* 21(10):1720-5.

9. Chu, K.C. (1998). Re: "Temporal trends in diabetes mortality among American Indians and Hispanics in New Mexico: birth cohort and period effects". *Am. J. Epidemiol.* 147(8):796-800.

10. Diabetes Care. (1995). Implication of treatment protocols in the Diabetes Control and Complications Trial. *Diabetes Care.* 18(3):361-76.

11. Feinglos, M.N. & Bethel, M.A. (1998). Treatment of type 2 diabetes mellitus. *Med. Clin. North Am.* 82(4):757-90.

12. Fischbach, F.T. (1995). *A Manual of Laboratory & Diagnostic Tests.* 5th Ed. Philadelphia. J.B. Lippincott Company.

13. Fonseca, V.A., Valiquett, T.R., Huang, S.M., Ghazzi, M.N., & Whitcomb, R.W. (1998). Troglitazone monotherapy improves glycemic control in patients with type 2 diabetes mellitus: a randomized, controlled study. The Troglitazone Study Group. *J. Clin. Endocrinol. Metab.* 83(9):3169-76.

14. Granberry, M.C., Schneider, E.F., & Fonseca, V.A. (1998). The role of troglitazone in treating the insulin resistance syndrome. *Pharmacotherapy.* 18(5):973-87.

15. Hanefeld, M. (1998). The role of acarbose in the treatment of non-insulin-dependent diabetes mellitus. *J. Diabetes Complications*. 12(4):228-37.

16. Hood, V.L., Kelly, B., Martinez, C., Shuman, S., & Secker-Walker, R. (1997). A Native American community initiative to prevent diabetes. *Ethn. Health*. 2(4):277-85.

17. Horton, E.S., Whitehouse, F., Ghazzi, M.N., Venable, T.C., & Whitcomb, R.W. (1998). Troglitazone in combination with sulfonylurea restores glycemic control in patients with type 2 diabetes. The Troglitazone Study Group. *Diabetes Care*. 21(9):1462-9.

18. Johnson, M.A. (1998). Medical nutrition therapy and combination medication for the treatment of type 2 diabetes. *Diabetes Care and Education*. 19(2):11-13.

19. Kappel, C. & Dills, D.G. (1998). Type 2 diabetes: update on therapy. *Compr. Ther*. 24(6-7):319-26.

20. Lam, K.S., Tiu, S.C., Tsang, M.W., Ip, T.P., & Tam, S.C. (1998). Acarbose in NIDDM patients with poor control on conventional oral agents. A 24-week placebo-controlled study. *Diabetes Care*. 21(7):1154-8.

21. Lee, A. & Morley, J.E. (1998). Metformin decreases food consumption and induces weight loss in subjects with obesity with type II non-insulin-dependent diabetes. *Obes. Res*. 6(1):47-53.

22. Mahoney, M.C. & Michalek, A.M. (1998). Health status of American Indians/Alaska Natives: genera; patterns of mortality. *Fam. Med*. 30(3):190-5.

23. Nursing 98 Books. (1998). *Drug Handbook*. Springhouse, Pennsylvania. Springhouse Corporation.

24. Paolisso, G., Amato, L., Eccellente, R., Gambardella, A., Tagliamonte, M.R., Varricchio, G., Carella, C., Giugliano, D., & D'Onofrio, F. (1998). Effect of metformin on food intake in obese subjects. *Eur. J. Clin. Invest*. 28(6):441-6.

25. Peterson, K.P., Pavlovich, J.G., Goldstrin, D., Little, R., England, J., & Peterson, C.M. (1998). What is hemoglobin A1c? An analysis of glycated hemoglobins by electrospray ionization mass spectrometry. *Clin. Chem*. 44(9):1951-8.

26. Prendergast, A. & Fulton, F.L. (1997). *Medical Terminology: A Text/Workbook*. 4th Ed.

27. Pronsky, Z.M. & Solomon, E. (1998). *Food-Medication Interactions*. 10th Ed. Phoenix, Arizona. Food-Medication Interactions, Publishers and Distributors.

28. Riddle, M.C. (1998). Learning to use troglitazone. *Diabetes Care*. 21(9):1389-90.

29. Robinson, A.C., Burke, J., Robinson, S., Johnston, D.G., & Elkeles, R.S. (1998). The effects of metformin on glycemic control and serum lipids in insulin-treated NIDDM patients with suboptimal metaboloc control. *Diabetes Care*. 21(5):701-5.

30. Roubicek, M., Vines, G.,& Gonzalez, S.A. (1998). Use of HbA1c in screening for diabetes. *Diabetes Care*. 21(9):1577-9.

31. Rusk, M.H. (1998). Effect of troglitazone in type 2 diabetes mellitus. *N.Engl. J. Med.* 339(6):406.

32. Scheen, A.J. & Lefebvre, P.J. (1998). Oral antidiabetic agents. A guide to selection. *Drugs.* 55(2):225-36.

33. Setter, S.M. (1998). New drug therapies for the treatment of diabetes. *Diabetes Care and Education.* 19(2):3-7.

34. Sharp, A.R. (1998). Nutritional implications of new medications to treat diabetes. *Diabetes Care and Education.* 19(2):8-10.

35. Swenson, K. (1998). Progression of care: Knowing when to change therapy for type 2 diabetes. *Diabetes Care and Education.* 19(2):14-16.

36. UK Prospective Diabetes Study Group. (1998). Effect of intensive blood-glucose control with metformin on complications in overweight patients with type 2 diabetes (UKPDS 34). UK Prospective Diabetes Study (UKPDS) Group. *Lancet.* 352(9131):854-65.

37. United Kingdom Prospective Diabetes Study Group. (1998). United Kingdom Prospective Diabetes Study 24: a 6-year, randomized, controlled trial comparing sulfonylurea, insulin, and metformin therapy in patients with newly diagnosed type 2 diabetes that could not be controlled with diet therapy. *Ann. Intern. Med.* 128(3):165-75.

38. Whitney, E.N., Cataldo, C.B., and Rolfes, S.R. (1998). *Understanding Normal and Clinical Nutrition*, 5th Ed. West/Wadsworth.

39. Wiener, K. & Roberts, N.B. (1998). The relative merits of haemoglobin A1c and fasting plasma glucose as first-line diagnostic tests for diabetes mellitus in nonpregnant subjects. *Diabet. Med.* 15(7):558-63.

40. Wolever, T.M., Chiasson, J.L., Josse, R.G., Hunt, J.A., Palason, C., Pedger, N.W., Ross, S.A., Ryan, E.A., and Tan, M.H., (1998). No relationship between carbohydrate intake and effect of acarbose on HbA1c or gastrointestinal symptoms in type 2 diabetic subjects consuming 30-60% of energy from carbohydrate. *Diabetes Care.* 21(10);1612-8.

CASE STUDY #29
DIABETIC COMPLICATIONS

INTRODUCTION

This is a complicated study involving diabetes, renal problems, cardiovascular problems, obesity, tube feedings, and a surgical procedure. It requires knowledge of several terms, involves practical application of nutrition counseling, and considers some habits of an ethnic group. Study all of the previously mentioned disease states before beginning the case.

SKILLS NEEDED

ABBREVIATIONS:

Knowledge of the following abbreviations is required in order to understand this case. You should learn these abbreviations before you begin to read the study.

BMI : body mass index
BS : bowel sounds
BUN : blood urea nitrogen
Cl : chloride
cl liqs : clear liquids
C/O : complains of
Cr : creatinine
D/C : discontinue
DVT : deep vein thrombosis
Dx : diagnosis
ER : emergency
g : gram
g/dl : grams per deciliter
GI : gastrointestinal tract
Glu : glucose
Hct : hematocrit
Hgb : hemoglobin
IDDM : insulin dependent diabetes mellitus
I.V. : intravenous
K : potassium
kg : kilogram
MCV: mean corpuscular volume

mEq/L : milliequivalent per liter
MH : medical history
MI : myocardial infarction
mg/dl : milligram per deciliter
mm^3 : cubic millimeter
Na : sodium
N/G : nasogastric
N/V : nausea and vomiting
P : phosphorous
PED : percutaneous endoscopic duodenostomy
po : by mouth
prn : as needed
PVD : peripheral vascular disease
RBW : reference body weight
R/O : rule out
SBO : small bowel obstruction
Ser Alb: serum albumin
S/P : status post
TF : tube feeding
WBC : white blood cell count
2° : secondary
2x : two times
μm^3: cubic microns

LABORATORY VALUES:

Look up the normal values for the following parameters: BUN, Cl, Cr, Glucose, Hct, Hgb, K, MCV, Na, P, Ser Alb, and WBC (Appendix B).

FORMULAS:

The formulas used in this case study include reference body weight, percent reference body weight, adjusted body weight, total energy and protein needs, and percentage calculations of tube feeding constituents based on flow rates (Appendices A, and D, Tables A - 1 through 13 and D - 1 and 5).

MEDICATIONS:
Become familiar with the following medications before reading the case study. Note the diet-drug interactions, dosages and methods of administration, gastrointestinal tract reactions, etc.

1. heparin; 2. insulin; 3. potassium supplements; 4. phosphorus supplements; 5. cisapride (Propulsid).

Mrs. M is a 64-year-old Cuban-American who was admitted to the ER with a Dx of DVT in her right leg and hyperglycemia, her fifth admission in the last year. She has a long standing MH that includes: Type I diabetes mellitus (formally IDDM), PVD, retinopathy, neuropathy, nephropathy, hypertension, and S/P MI. Mrs. M is 5'3" and weighs 252 lbs. She lives with her son, who is also obese, and does not understand the importance of diet. Because of her problems, her son feels sorry for her and goes along with whatever she wants. This solicitousness is usually centered around eating. Her son works as a short-order cook and enjoys making dishes for her that she likes. They live in Miami, where there is a large Cuban population and where many Cuban foods are avilable. Some of the foods Mrs. M and her son eat on a daily basis are fried plantains, dried black beans, and chick peas. Rice is always eaten with legumes. Arroz con qui is another favorite dish. Sweet potatoes and yams (yuca) are eaten more often than white potatoes, but french fries are consumed often. Chicken and pork are more frequent choices than beef. Their favorite vegetables include fried eggplant, beets, and greens. The vegetables are cooked with salt pork, ham, or lard. Mrs. M and her son drink several cups of strong coffee per day with sugar. Sugar is also used in cooking, for instance, the plantains are fried in a skillet with a little oil and then sprinkled with sugar and sugar is sometimes added to the sweet potatoes. Very little fruit is eaten, although they do drink orange juice on a regular basis. They also eat many foods not common to the Cuban culture.

Mrs. M's eyesight is poor, but she is not blind. She enjoys watching her son bowl at the local bowling alley two to three times a week. Whenever her son takes her to the bowling alley he buys her a large hamburger, french fries, and a large soda. She knows it is not good for her, but sometimes she will drink a couple of beers while she is watching.

Mrs. M's kidney function is not seriously abnormal but it has been affected by her weight and her hypertension. Nephrotic syndrome with the concurrent proteinuria and edema is slight, but her nephrologist is concerned that it will become much worse if she does not start following her diet. Mrs. M had a slight MI one year ago 2° to atherosclerosis. Angioplasty was successful, and again she was warned that she needed to change her diet or she could soon have more severe blockages. A cardiologist explained to her that she would eventually require open heart surgery if she did not lose weight and follow her diet. All the emphasis on diet was ignored. After her MI, Mrs. M gained 40 lbs due to the decreased activity. The increased weight caused more inactivity, and hence, a DVT.
****

QUESTIONS:

1.   Define the following as they relate to diabetes:

      Hyperglycemia:

Retinopathy:

Neuropathy:

Nephropathy:

Nephrotic Syndrome:

Proteinuria:

Edema:

Atherosclerosis:

Angioplasty:

Nephrologist:

Cardiologist:

Myocardial Infarction:

Hypertension:

2.   Determine Mrs. M's RBW and percent of RBW (Appendix A, Tables A - 7 through 10).

3.   Calculate Mrs. M's BMI (Appendix A, Tables A 12 & 13).

4   Give the pathophysiology of the following:

Retinopathy:

Neuropathy:

Nephropathy:

Nephrotic Syndrome:

5.   Describe the following foods:

Plantains:

Yuca:

Chick Peas:

Yams:

Legumes:

Arroz con qui:

6.   Some of Mrs. M's food choices are extremely poor for her medical condition. Her intake is complicated by her obese son who cooks for her. The diet she should be following is complex because of the multiple problems she has. Each of Mrs. M's problems is listed as a heading below. Under each heading list the foods mentioned in the case study that Mrs. M should avoid. Many foods may be listed more than once.

OBESITY   **DIABETES**   **RENAL**   **CARDIOVASCULAR**

7.   For the above mentioned foods that should be avoided, suggest an appropriate substitute.

8. While Mrs. M is hospitalized for DVT, the RD will have a chance to work with her. Outline the steps that you, as the RD, would take to teach Mrs. M her diet and the importance of following it.

9. Considering the lifestyle presented, what behavioral changes would you suggest?

****

Mrs. M's lab values were as follows:

TEST	RESULT	NORM	TEST	RESULT	NORM	TEST	RESULT	NORM
Hgb	13.0 g/dl	12 - 16	MCV	84 µm^3	82 - 98	Glu	198 mg/dl	65 - 110
Hct	38%	36 - 48	WBC	7.3x10^3 /mm^3	5 - 10 x 10^3	Na	144 mEq/L	135 - 145
K	3.1 mEq/L	3.5 - 5.3	Cl	100 mEq/L	98 - 106	Ser Alb	3.3 g/dl	3.9 - 5.0
BUN	25 mg/dl	7 - 18	Cr	1.4 mg/dl	0.6 - 1.3	P	4.4 mg/dl	2.5 - 4.5

****

QUESTIONS CONTINUED:

10. Mrs. M hyperglycemia and nephrotic syndrome. How are these conditions going to affect her lab values?

****

Mrs. M was treated with I.V. heparin therapy, insulin, potassium and phosphorus supplementation prn, bed rest, and a 1000 kcal, 2 g Na diet, with a protein intake not to exceed .7 g per kg RBW.

****

11.   Calculate Mrs. M's adjusted body weight (Appendix A, Table A - 11).

12.   Why would the MD order a protein restriction of .7 g/kg of RBW? Explain.

13.   Why use the RBW weight instead of the adjusted body weight? In your answer, relate how this would affect her protein requirement.

14.   Why was it important for Mrs. M to receive potassium and phosphorous I.V. along with insulin?

****

Mrs. M progressed well on her treatment and the clot resolved. The RNs started getting her out of bed and ambulating her 2x daily. They were preparing her for D/C when she developed a new symptom. She C/O not getting enough to eat most of the time, but one day she refused her tray. She said she was still full from the last meal. Later she felt nauseous and began vomiting. She continued with N/V to such a degree that a N/G tube had to be placed. Her abdomen became distended and hard to touch. Her BS were decreased. Mrs. M had an ileus. The physicians had to R/O a SBO. First an esophagogastroduodenoscopy was done and the results were negative.

Gastric emptying time was studied and a significant delay was found. Venography studies indicated that ischemia of the gastric arteries was slowing down the blood supply to the stomach and causing a decrease in gastric functioning. This was termed gastratroparesis 2° to diabetic gastrovasculitis.

The physicians were not sure if this was a permanent condition or if it would improve enough for Mrs. M to be able to eat again. The GI tract seemed to be functioning well beyond the stomach. All Mrs. M could tolerate po was cl liqs. Therefore, until Mrs. M recovered from this setback, a PED was performed and a feeding tube was placed. The MD also prescribed cisapride to aid in gastric emptying when po feedings were resumed.
****

QUESTIONS CONTINUED:

15. Define the following terms:

    Ileus:

    Venography:

    Ischemia:

    Esophagogastroduodenoscopy:

    Gastratroparesis:

    Gastrovasculitis:

16. Summarize what has happened to Mrs. M with this latest diabetic complication and explain what may have caused this.

17. What is the action of cisapride and what are its side effects?

18. Describe the placement and purpose of a PED.

19. Considering all of the problems Mrs. M has, what TF would you recommend? Justify your answer.

20. Describe the initial strength and flow rate you would use, the progression to the final flow rate, and the total kcals and protein Mrs. M would be receiving at the final flow rate (in total kcals and total grams and in kcals and grams per kg of RBW).

‣ ADDITIONAL OPTIONAL QUESTION ‣

Tube Feeding Drill:

21. Using the table below, compare several of the enteral nutritional supplements that would be appropriate for a diabetic with Mrs. M's complications (Appendix F).

| Product | Producer | Form | Cal/ml | Non-pro Cal/g N | g/L | | | Na mg | K mg | mOsm/kg Water | Vol to meet RDAs in ml | g of fiber /L | Free water /L in ml |
					Pro	CHO	Fat						

Related References

1. American Diabetes Association. (1993). Implications of the Diabetes Control and Complications Trial. American Diabetes Association. *Diabetes.* 42(11):1555-8.

2. American Diabetes Association. (1994). Nutrition recommendations and principles for people with diabetes mellitus. *J. Am. Diet. Assoc.* 94:504.

3. Brown, S.L., Pope, J.F., Hunt, A.E., & Tolman, N.M. (1998). Motivational strategies used by dietitians to counsel individuals with diabetes. *Diabetes Educ.* 24(3):313-8.

4. Caudle, P. (1993). Providing culturally sensitive health care to Hispanic clients. *Nurse Pract.* 18(12):40, 43-6, 50-1.

5. Chang, C.S., Lien, H.C., Yeh, H.Z., Poon, S.K., Tung, C.F., & Chen, G.H. (1998). Effect of cisapride on gastric dysrhythmia and emptying of indigestible solids in type-II diabetic patients. *Scand. J. Gastroenterol.* 33(6):600-4.

6. Diabetes Care. (1995). Implication of treatment protocols in the Diabetes Control and Complications Trial. *Diabetes Care.* 18(3):361-76.

7. Enck, P. & Frieling, T. (1997). Pathophysiology of diabetic gastroparesis. *Diabetes.* Suppl2:S77-81.

8. Fischbach, F.T. (1995). *A Manual of Laboratory & Diagnostic Tests.* 5th Ed. Philadelphia. J.B. Lippincott Company.

9. Flegal, K.M., Ezzati, T.M., Harris, M.I., Haynes, S.G., Juarez, R.Z., Knowler, W.C., Pereztable, E.J., & Stern, M.P. (1991). Prevalence of diabetes in Mexican Americans, Cubans, and Puerto Ricans from the Hispanic Health and Nutrition Examination Survey, 1982-1984. *Diabetes Care.* 14(7):628-38.

10. Hadden, D.R. (1994). The Diabetes Control and Complications Trial (DCCT): What every endocrinologist needs to know. *Clin. Endocrinol.* 40(3):293-4.

11. Hanis, C.L., Hewett-Emmett, D., Bertin, T.K., & Schull, W.J. (1991). Origins of U.S. Hispanics. Implications for diabetes. *Diabetes Care.* 14(7):618-27.

12. Hentzen, D.H. (1994). From the president: the results of the Diabetes Control and Complications Trial. *Diabetes Educ.* 20(2):103.

13. James, W.P. (1998). What are the health risks? The medical consequences of obesity and its health risks. *Exp. Clin. Endocrinol. Diabetes.* 106 Suppl 2:1-6.

14. Khan, L.K., Sobal, J., & Martorell, R. (1997). Acculturation, socioeconomic status, and obesity in Mexican Americans, Cuban Americans, and Puerto Ricans. *Int. J. Obes. Relat. Metab. Disord.* 21(2):91-6.

15. Kim, C.H. & Nelson, D.K. (1998). Venting percutaneous gastrostomy in the treatment of refractory idiopathic gastroparesis. *Gastrointest. Endosc.* 47(1):67-70.

16. Kitter, P.G. & Sucher, K.P. (1998). *Food and Culture in America*. 2nd Ed. Belmont, CA. West/Wadsworth.

17. Nursing 98 Books. (1998). *Drug Handbook*. Springhouse, Pennsylvania. Springhouse Corporation.

18. Pawson, I.G., Martorell, R., & Mendoza, F.E. (1991). Prevalence of overweight and obesity in US Hispanic populations. *Am. J. Clin. Nutr.* 53(6 Suppl):1522S-1528S.

19. Prendergast, A. & Fulton, F.L. (1997). *Medical Terminology: A Text/Workbook*. 4th Ed.

20. Pronsky, Z.M. & Solomon, E. (1998). *Food-Medication Interactions*. 10th Ed. Phoenix, Arizona. Food-Medication Interactions, Publishers and Distributors.

21. Whitney, E.N., Cataldo, C.B., and Rolfes, S.R. (1998). *Understanding Normal and Clinical Nutrition*, 5th Ed. West/Wadsworth.

CASE STUDY #30
CLOSED HEAD INJURY

INTRODUCTION

This is part one of a two-part study about a closed head injury. This is a good introduction to the use of tube feedings for unconscious patients. It requires a basic knowledge of starting tube feedings and the importance of monitoring tube feedings as the patient's condition improves.

SKILLS NEEDED

ABBREVIATIONS:

Knowledge of the following abbreviations is required in order to understand this case. You should learn these abbreviations before you begin to read the study.

BMR : basal metabolic rate
cc/hr : cubic centimeters per hour
CHI : closed head injury
CO_2 : carbon dioxide
ER : emergency room
Hct : hematocrit
Hgb : hemoglobin

ICP : intracranial pressure
I.V. : intravenous
qid : four times a day
MVA : motor vehicle accident
NSICU : neurosurgical intensive care unit
PEG : percutaneous endoscopic gastrostomy
RBW : reference body weight
YOWM : year old white male

LABORATORY VALUES:

You will need to be able to interpret the nutritional significance of the following laboratory values for this case study: Hct, Hgb, and ICP.

FORMULAS:

The formulas used in this case study include reference body weight and percent reference body weight, total caloric needs using the Harris-Benedict equation and appropriate stress factors, total protein needs, and calculation of tube feeding flow rates. The formulas can be found in Appendices A and D, Tables A - 7 through 10 and D - 1 and 5.

RK is a 25 YOWM who was in a MVA nine months ago. He was thrown from his vehicle and received multiple fractures, contusions, and a closed head injury (CHI). The CHI was severe and RK stayed in a neurosurgical intensive care unit for five weeks. He was unconscious most of the time and had to be fed via a nasogastric feeding tube. He received normal saline and electrolytes via I.V. It was necessary to monitor his intracranial pressure, so a ventriculostomy cather was put in place. His ICP remained in the 30s for most of the first week and gradually returned to normal. RK's lab values were normal upon admission with the exception of Hgb and Hct, which were low due to bleeding. Packed cells were administered and RK equilibrated rapidly.

After one week in NSICU, RK began to respond to physical stimuli but not verbal stimuli. In the third week, RK opened his eyes and started responding to sound but still would not obey verbal commands. By the fourth week, he was moving all limbs well but without coordination. At the beginning of the fifth week RK was transferred to a neurosurgical ward and continued to be monitored. He was still receiving a tube feeding but the screw had been removed since his ICP returned to normal.

According to his brother, RK was 5'11" and weighed 180 lbs at the time of the accident. He was not able to be weighed in the ER when admitted. When RK was transferred to a ward, he weighed 135 lbs. While still in NSICU, a feeding tube was placed in the small bowel and he received one-half-strength Ensure Plus at 30 cc/hr. This was started five days after admission. After one day, the tube feeding was changed to full-strength and gradually increased to 50 cc/hr. The neurosurgeon attending to RK was concerned about fluid overload and CO_2 buildup secondary to feeding. He would not even consider starting a tube feeding prior to 72 hours post injury. After being on Ensure Plus for a week, the physician changed to Two Cal HN for a more concentrated feeding. The Two Cal HN was started at one-half-strength at a rate of 30 cc/hr. After 24 hours the feeding was changed to full-strength. It was gradually increased to 50 cc/hr. RK tolerated this feeding for his entire stay in the hospital. After he was transferred to the ward, RK received a PEG and was continued on Two Cal HN but by bolus instead of continuous drip. The physician ordered the bolus feedings to be given at full strength in a quantity sufficient to equal the amount given in 24 hours by the continuous drip. The end of the feeding tube was in the stomach.

With RK responding positively to the PEG and maintaining a normal blood pressure and vital signs, he was not monitored as closely as he was during his stay in NSICU. This was particularly true for his weight. Immediately after a CHI, most patients respond with a very high basal metabolic rate and lose weight rapidly. The nutritional treatment for this is high caloric/high protein intake. After recovery, the BMR returns to normal or lower than normal and weight gain begins at a rapid pace. Many victims of CHIs are bed ridden and are unable to exercise, thus requiring fewer kilocalories. This results in excessive weight gain as adipose tissue if the tube feeding is not decreased. This was the case with RK. His mental condition never improved. He still responded to painful stimuli and to sound but could not follow any commands. He was able to move all four limbs on his own but without any coordination and not in response to commands. Posturing of all limbs was moderately severe. He did not appear to respond any differently to the voices of his family members than he did to the voices of the hospital staff. His physician and his parents were pleased to see him gaining weight but did not realize how heavy he was becoming. His tube feeding was reduced to 40 cc/hr.
****

QUESTIONS:

1.   Define closed head injury.

2.   Determine RK's reference body weight.

3.   Calculate RK's caloric needs at admission and when he was transferred to a ward.

4.   Calculate RK's protein needs at admission and when he was transferred to a ward.

5.   Calculate the kilocalories and protein RK was receiving from Ensure Plus at 35cc/h half-strength and at 50cc/hr full-strength.

6.   Compare Ensure Plus and Two Cal HN for osmolarity, kcals/cc, protein/L, and free water.

7.   How many kcals did RK receive with Two Cal HN full-strength at 50 cc/hr?

8.   What was the reason for starting the Ensure Plus at half-strength?

9.   Would you still start at half-strength if the tip of the feeding tube was in the stomach instead of the small bowel? If not, explain why not.

10.   What was the reason for starting the Two Cal HN at half-strength?

11.   Compare bolus feeding to a continuous drip as to expense, and advantages and disadvantages for the patient and for the nurses.

12.   Calculate the amount of Two Cal HN per bolus feeding, and the frequency of feedings necessary to equal the amount given in 24 hours as a continuous feeding at 50 cc/h. Plan to give the bolus feedings between 7 A.M. and 10 P.M. How much would you reduce each bolus feeding to equal a reduction in the continuous feeding from 50 cc/hr to 40 cc/hr?

13.   What is a PEG?

14.   Why would this be used instead of a nasogastric tube?

15.   What is a ventriculostomy and what does it have to do with a CHI?

16.   What relationship does CO_2 have with ICP and feeding rate?

17.   Why would the neurosurgeon not even consider a tube feeding until 72 hours post injury?

****

RK was sent home with the following nutritional orders:

▸ Two Cal HN bolus feeding via PEG.
▸ Feed full-strength 240 cc qid.
▸ Flush tube with 30 cc of water after each feeding.

****

18.   How many kcals and protein and how much free water does this provide?

19.   What would be the equivalent flow rate per hour?

20.   How much total free water is he receiving with the water being used to flush the tube after feeding?

21.   What is the rule of thumb water requirement for an adult receiving a tube feeding?

▸ ADDITIONAL OPTIONAL QUESTION ◂

Tube Feeding Drill:

22.   Using the table below, compare several of the enteral nutritional supplements that are formulated to help reduce CO_2 production.

Product	Producer	Form	Cal/ml	Non-pro Cal/g N	g/L			Na mg	K mg	mOsm /kg Water	Vol to meet RDAs in ml	g of fiber /L	Free water /L in ml
					Pro	CHO	Fat						

Related References

1.   Bosscha, K., Nieuwenhuijs, V.B., Vos, A., Samsom, M., Roelofs, J.M., & Akkermans, L.M. (1998). Gastrointestinal motility and gastric tube feeding in mechanically ventilated patients. *Crit. Care Med.* 26(9):1510-7.

2.   Fertl, E., Steinhoff, N., Schofl, R., Potzi, R., Doppelbauer, A., Muller, C., & Auff, E. (1998). Transient and long-term feeding by means of percutaneous endoscopic gastrostomy in neurological rehabilitation. *Eur. Neurol.* 40(1):27-30.

3.   Fischbach, F.T. (1995). *A Manual of Laboratory & Diagnostic Tests.* 5th Ed. Philadelphia. J.B. Lippincott Company.

4.   Loan, T., Magnuson, B., & Williams, S. (1998). Debunking six myths about enteral feeding. *Nursing.* 28(8):43-9.

5.   Moore, R., Najarian, M.P., & Konvolinka, C.W. (1989). Measured energy expenditure in severe head trauma. *J. Trauma.* 29(12):1633-6.

6.   Nursing 98 Books. (1998). *Drug Handbook.* Springhouse, Pennsylvania. Springhouse Corporation.

7.   Prendergast, A. & Fulton, F.L. (1997). *Medical Terminology: A Text/Workbook.* 4th Ed.

8.   Pronsky, Z.M. & Solomon, E. (1998). *Food-Medication Interactions.* 10th Ed. Phoenix, Arizona. Food-Medication Interactions, Publishers and Distributors.

9.   Sacks, G.S., Brown, R.O., Teague, D., Dickerson, R.N., Tolley, E.A., & Kudsk, K.A. (1995). Early nutrition support modifies immune function in patients sustaining severe head injury. *JPEN.* 19(5):387-92.

10.   Thelan, L.A., Davie, J.K., Urden, L.D., & Lough, M.E. (1994). *Critical Care Nursing. Diagnosis and Management.* 2nd Ed. St. Louis, MO. Mosby.

11.   Young, B., Ott, L., Kasarskis, E., Rapp, R., Moles, K., Dempsey, R.J., Tibbs, P.A., Kryscio, R., & McClain, C. (1996). Zinc supplementation is associated with improved neurologic recovery rate and visceral protein levels of patients with severe closed head injury. *J. Neurotrauma.* 13(1):25-34.

12.   Whitney, E.N., Cataldo, C.B., & Rolfes, S.R. (1998). *Understanding Normal and Clinical Nutrition,* 5th Ed. West/Wadsworth.

CASE STUDY #31
HOME HEALTH CARE

INTRODUCTION
This is part two of a study of closed head injuries. In this study, the patient is discharged from the hospital but still requires nursing care and nutritional assessment. This case provides an introduction to the use of tube feedings at home as well as home health care consulting.
SKILLS NEEDED

ABBREVIATIONS:
Knowledge of the following abbreviations is required in order to understand this case. You should learn these abbreviations before you begin to read the study.

bid : twice a day	mm^3 : cubic millimeter
Ca : calcium	mg/dl : milligrams per deciliter
cc : cubic centimeters	NSICU : neurosurgical intensive care unit
D/C : discontinued	PEG : percutaneous endoscopic gastrostomy
g/dl : grams per deciliter	prn : as needed
Hct : hematocrit	qid : four times a day
Hgb : hemoglobin	q4h : every four hours
kg : kilogram	WBC : white blood cell count
LBM : lean body mass	μm^3 : cubic microns
MCV : mean corpuscular volume	

LABORATORY VALUES:
The normal values for the following parameters will be needed for this case study: Hct, Hgb, MCV, BUN, Cr, and glucose.

FORMULAS:
The formulas used in this case study include total caloric needs using the Harris-Benedict equation and appropriate stress factors, total protein needs, and calculation of bolus tube feeding rates. The formulas can be found in Appendix D, Table D - 1, 2, and 5.

RK continued to do well in the hospital from a medical standpoint but did not improve mentally. RK was awake and alert and responded to sound by looking in the direction of the sound but could not obey any commands. The nurses responsible for RK could not tell any difference in his response to family as compared to strangers, yet, his mother insisted that he responded more to her voice than anyone else's. RK's movements were totally spastic and without any coordination. Posturing of both arms and legs was still evident although it was not as bad as previously. The physician did not give the family any hope that RK would improve mentally. As long as someone fed him and kept him clean, he would live a long but unresponsive life.

His mother refused to accept this prognosis. She insisted that he could understand her and was responding to her commands and not just to her voice. No one else shared this opinion. This is frequently the case with a mother and a nonresponsive patient. The physician wanted to discharge RK to a nursing home but his mother insisted on taking him home. She did not work and vowed to become his nurse. The physician finally agreed. It was necessary for the physician to D/C RK with orders so he asked the attending nurse how he was doing on the hospital orders. The response was "fine" so he ordered her to D/C him to home with the same orders. The nurse failed to tell him that RK's weight was up to 160 lbs. His discharge orders were as follows:

- ‣ 1. Continue Two Cal HN bolus feeding via PEG.
- ‣ 2. Feed full strength 240 cc qid.
- ‣ 3. Flush tube with 30 cc of water after each feeding.
- ‣ 4. Instruct family how to feed patient.

****

QUESTIONS:

1.   Define the following terms:

      prognosis:

      posturing:

2.   What does posturing indicate?

3.   Would posturing affect RK's caloric needs? Why or why not?

4.   Compare RK's weight before the accident, at his discharge from NSICU, and upon his discharge from the hospital. Compare the probable composition of his body prior to the accident and at the time of discharge from the hospital (fat to lean body mass ratio).

5.   Calculate RK's caloric and protein requirements at discharge from the hospital. Compare this to the amount of calories and protein he is receiving from Two Cal HN at 240 cc qid.

6.   What suggestions would you have at this point concerning his nutrition? Consider such factors as inactivity, Ca, trace minerals, fluid, and fiber.

7.   What is the purpose of flushing the tube with water after each feeding?

****

RK was discharged home and his mother learned very quickly how to give excellent nursing care. She followed the instructions she received from the nurses exactly. RK continued to tolerate the bolus feedings as far as she could tell but getting up during the night to feed every four hours was really a strain. Her husband and her son helped her to accomplish this, but after a few weeks they were all very tired. RK continued to gain weight. His mother had no way of weighing him but she could tell he was getting considerably heavier and very difficult to move. She did not mention this to the home health care nurse because she assumed that it meant he was doing well. After a short time RK developed another problem: he became constipated. Several days passed without a bowel movement so RK's mother called his physician for help. He sent a lab tech to draw blood for some basic tests and found the following:

TEST	RESULT	NORM	TEST	RESULT	NORM	TEST	RESULT	NORM
Hgb	21.0 g/dl	14 - 17	MCV	87 μm^3	82 - 98	Glu-cose	145 mg/dl	65 - 110
Hct	54%	42 - 52	BUN	25 mg/dl	7 - 18	Cr	0.9 mg/dl	0.6 - 1.3

He then asked her home health care agency to send a dietitian for an assessment. The agency contacted one of their consultants and sent an RD to RK's house.

After assessing the patient and reviewing RK's old hospital records and current lab values, the RD found RK to be dehydrated and made the following suggestions:

- ▸ 1. Change RK's feeding to Enrich, 350 cc qid during wakening hours.
- ▸ 2. Flush feeding tube with 30 cc water before and after each feeding.
- ▸ 3. Give 60 cc of prune juice bid via PEG.
- ▸ 4. Obtain bed weight and monitor weight monthly and every three months after weight stabilizes.
- ▸ 5. RD to assess patient every two weeks until weight stabilizes, then prn.

****

QUESTIONS CONTINUED:

8. What lab values indicated dehydration?

9. Why was RK dehydrated?

10. Compare the composition of Enrich and Two Cal HN and explain the reasoning behind using Enrich instead of Two Cal HN.

11. How much free water should RK have been receiving with Two Cal HN and how much should he receive with Enrich?

12. Explain the physiology of his constipation and list all of the contributing factors. Include in your explanation how a feeding with such a high osmolarity, as Two Cal HN, could cause constipation.

13. How will his new regimen prevent constipation?

14. What is the purpose of prune juice bid?

15. After a bed weight is obtained the RD will be able to calculate a specific caloric and protein need. Considering RK's lack of activity and posturing, estimate his kcal and protein needs per kg of body weight.

16. RK's lack of activity and excessive intake is causing his lean body mass to decrease and his adipose tissue to increase. His LBM to fat ratio is not going to be like that of a person who is active. Suggest a technique the RD could use to determine RK's actual body composition.

► ADDITIONAL OPTIONAL QUESTIONS ◄

Tube Feeding Drill:

17.   Compare Two Cal with at least three other tube feedings that provide 2.0 kcals per cc using the following table (there is room for seven comparisons).

Product	Producer	Form	Cal/ml	Non-pro Cal/g N	g/L			Na mg	K mg	mOs m/kg Water	Vol to meet RDAs in ml	g of fiber /L	Free water /L in ml
					Pro	CHO	Fat						

18.   Compare Enrich with at least five tube feedings that have fiber as part of their formula (there is room for seven comparisions).

Product	Producer	Form	Cal/ml	Non-pro Cal/g N	g/L			Na mg	K mg	mOs m/kg Water	Vol to meet RDAs in ml	g of fiber /L	Free water /L in ml
					Pro	CHO	Fat						

Related References

1. Fertl, E., Steinhoff, N., Schofl, R., Potzi, R., Doppelbauer, A., Muller, C., & Auff, E. (1998). Transient and long-term feeding by means of percutaneous endoscopic gastrostomy in neurological rehabilitation. *Eur. Neurol.* 40(1):27-30.

2. Fischbach, F.T. (1995). *A Manual of Laboratory & Diagnostic Tests.* 5th Ed. Philadelphia. J.B. Lippincott Company.

3. Goff, K. (1998). Enteral and parenteral nutrition transitioning from hospital to home. *Nurs. Case Manag.* 3(2):67-74.

4. Loan, T., Magnuson, B., & Williams, S. (1998). Debunking six myths about enteral feeding. *Nursing.* 28(8):43-9.

5. Nursing 98 Books. (1998). *Drug Handbook.* Springhouse, Pennsylvania. Springhouse Corporation.

6. Prendergast, A. & Fulton, F.L. (1997). *Medical Terminology: A Text/Workbook.* 4th Ed.

7. Pronsky, Z.M. & Solomon, E. (1998). *Food-Medication Interactions.* 10th Ed. Phoenix, Arizona. Food-Medication Interactions, Publishers and Distributors.

8. Whitney, E.N., Cataldo, C.B., & Rolfes, S.R. (1998). *Understanding Normal and Clinical Nutrition,* 5th Ed. West/Wadsworth.

9. Williams, D.M. (1998). The current state of home nutrition support in the United States. *Nutrition.* 14(4):416-9.

Related References

1. Addolorato, G., Capristo, E., Greco, A.V., Stefanini, G.F., & Gasbarrini, G. (1997). Energy expenditure, substrate exidation, and body composition in subjects with chronic alcoholism: new finding from metabolic assessment. *Alcohol Clin. Exp. Res.* 21(6):962-7.

2. Addolorato, G. Capristo, E., Stefanini, G.F., & Gasbarrini, G. (1998). Metabolic features and nutritional status in chronic alcoholics. *Am. J. Gastroenterol.* 93(4):555-6.

3. Apte, M., Norton, I., Haaber, P., Applegate T., Kesten, M., McCaughn, Pirola, R., & Wilson, J. (1998). The effect of alcohol on pancreatic enzymes dietary artifact? *Biochem. Biophys. Acta.* 2;1379(3):314-24

4. Cabre, E. & Gassull, M.A., (1994). Nutritional therapy in liver disease. *Acta. Gastroenterol. Belg.* 57(1):1-12.

5. Fischbach, F.T. (1995). *A Manual of Laboratory & Diagnostic Tests.* 5th Ed. Philadelphia. J.B. Lippincott Company.

6. Gloria, L., Cravo, M., Camilo, M.E., Resende, M., Cardoso, J.N., Oliverira, A.G., Leitao, C.N., & Mira, F.C. Nutritional deficiencies in chronic alcoholics: relation to dietary intake and alcohol consumption. *Am. J. Gastroenterol.* 92(3):485-9.

7. Hill, D.B. & Kugelmas, M. (1998). Alcoholic liver disease. Treatment strategies for the potentially reversible stages. *Postgrad. Med.* 103(4):261-4, 267-8, 273-5.

8. Lieber, C.S., & Leo, M.A. (1998). Metabolism of ethanol and some associated adverse effects on the liver and stomach. *Recent Dev. Alcohol.* 14:7-40.

9. Marsano, L. & McClain, C.J. (1991). Nutrition and alcoholic liver disease. *J. Parenter. Enteral Nutr.* 15(3):337-44.

10. McClave, S.A., Snider, H., Owens, N., & Sexton, L.K. (1997). Clinical nutrition in pancreatitis. *Dig. Dis. Sci.* 42(10):2035-44.

11. McClave, S.A.,Spain, D.A., & Snider, H.L. (1998). Nutritional management in acute and chronic pancreatitis. *Gastroenterol. Clin. North Am.* 27(2):421-34.

12. Nursing 98 Books. (1998). *Drug Handbook.* Springhouse, Pennsylvania. Springhouse Corporation.

13. Pennington, C.R. (1998). Feeding the inflamed pancreas. *Gut.* 42(3):315-6.

14. Prendergast, A. & Fulton, F.L. (1997). *Medical Terminology: A Text/Workbook.* 4th Ed.

15. Pronsky, Z.M. & Solomon, E. (1998). *Food-Medication Interactions.* 10th Ed. Phoenix, Arizona. Food-Medication Interactions, Publishers and Distributors.

16. Sarin, S.K., Dhingra, N., Bansal, A., Malhotra, S., & Guptan, R.C. (1997). Dietary and nutritional abnormalities in alcoholic liver disease: a comparison with chronic alcoholics without liver disease. *Am. J. Gastroenterol.* 92(5):777-83.

17. Whitney, E.N., Cataldo, C.B., & Rolfes, S.R. (1998). *Understanding Normal and Clinical Nutrition*, 5th Ed. West/Wadsworth.

18. Windsor, A.C., Kanwar, S., Li, A.G., Barnes, E., Guthrie, J.A., Spark, J.I., Welsh, F., Guillou, P.J., & Reynolds, J.V. (1998). Compared with parenteral nutrition, enteral feeding attenuates the acute phase response and improves disease severity in acute pancreatitis. *Gut.* 42(3):431-5.

CASE STUDY #33
ALCOHOLIC CIRRHOSIS

INTRODUCTION

This is the second study in a series of two concerned with liver disease. Mr. N was being treated for Laennec's cirrhosis by diet and medication. As his condition worsened, a tube feeding became necessary. This will involve a more detailed study of cirrhosis, some additional terms, and some knowledge of tube feedings for hepatic disease.

SKILLS NEEDED

ABBREVIATIONS:

Knowledge of the following abbreviations is required in order to understand this case. You should learn these abbreviations before you begin to read the study.

ALT : alanine aminotransferase

AST : aspartate aminotransferase

BRB : bright red blood

cc/h : cubic centimeters per hour

CHO : carbohydrate

DTs : delirium tremens

D_5W : 5 % dextrose in water

Dx : diagnosis

g : gram

g/dl : gram per deciliter

GGTP : gamaglutamyl transpeptidase

I.V. : intravenous

Na : sodium

N/G : nasogastric

NH_3 : ammonia

NPO : nothing by mouth

sec : seconds

TF : tube feeding

μmol/L : micromole per liter

2x : two times

LABORATORY VALUES:

The normal values for the following parameters will be needed for this case study: GGTP, Serum Albumin, ALT, AST, and NH_3 (Appendix B).

FORMULAS:

The formulas used in this case study include: basal energy expenditure using the Harris-Benedict equation, total energy expenditure using an appropriate stress factor, protein needs, simple percentages, and tube feeding and I.V. flow rates (Appendix D, Tables D - 1 and 5)

MEDICATIONS:

Become familiar with the following medications before reading the case study. Note the diet-drug interactions, dosages and methods of administration, gastrointestinal tract reactions, etc.

1. spironolactone (Aldactone); 2. furosemide (Lasix); 3. lactulose enema; 4. Neomycin

Mr. N's condition seemed to worsen. He developed ascites and pedal edema. His urinary output was decreasing. His diet consisted of the energy and protein levels recommended by the dietitian and a 1 g Na restriction. Mr. N did not like his diet at all. He ate sparingly and was not receiving the nutrients he needed. He continued to lose weight. After a few days without alcohol, he began to have DTs and had to be restrained. During this time he began to hallucinate and use very abusive language. He also started to exhibit asterixis. Blood NH_3 was drawn and was 25 μmol/L. His intake dropped to almost nothing. He was receiving only D_5W by I.V. with added vitamins, minerals, and electrolytes. That night he started throwing up large amounts of BRB. A Sengstaken-Blakemore tube had to be placed. Mr. N bled so excessively that whole blood had to be administered. His physician added to his Dx the following:

1. hepatic encephalopathy
2. portal hypertension
3. esophageal varices

Mr. N was now in a semi-comatose state and had to keep the Sengstaken-Blakemore tube in place for another day. After the tube was removed, Mr. N had a N/G tube in place to low suction. He was NPO and was receiving D_5W by I.V. with electrolytes, vitamins, and minerals. Mr. N continued to have severe ascites and pedal edema with reduced urinary output. He had some blood drawn again and his NH_3 was now up to 92 μmol/L. His prothrombin time was off by 4 sec. His serum albumin was down to 2.2 g/dl. AST was 2x ALT and GGTP was 800. New orders for Mr. N included the following:

1. spironolactone (Aldactone) I.V.
2. furosemide (Lasix) I.V.
3. lactulose enema
4. Neomycin via N/G tube
5. D_5W at 75 cc/hr

The physician called for the dietitian to recommend an appropriate tube feeding and flow rate.
****

QUESTIONS:

1.   What is the mechanism of action of the following drugs:

       Spironolactone (Aldactone):

       Furosemide (Lasix):

2.   What are the nutritional implications of these drugs, especially as they pertain to Mr. N's condition?

3.   Explain the mechanism of action of lactulose.

4.   Explain the mechanism of action of Neomycin.

5.   Define the following terms:

      Portal hypertension:

      Esophageal varices:

      Sengstaken-Blakemore tube:

      Hepatic encephalopathy:

      Asterixis:

6.   Explain the pathophysiology of esophageal varices and portal hypertension as it relates to
      liver disease and Mr. N's bleeding.

7.   What does it mean to have a N/G tube to low intermittent suction?

8.   What is the relationship between prothrombin time and liver disease?

9. Mr. N's AST was 2x ALT and GGTP was 800. What does this tell the physician concerning Mr. N's liver?

10. What is the significance of Mr. N's NH_3 level being elevated prior to his bleeding? What is the significance of it being elevated to an even greater degree after his bleeding?

11. Before his bleeding, what diet would have been appropriate for Mr. N? Be specific for protein, carbohydrate, fat, total energy, Na, fluid, and vitamins and minerals.

12. The tube feeding Mr. N should receive is obviously one designed for liver failure. List the characteristics of a hepatic TF and explain the reasoning behind each characteristic.

13. Are there any vitamins or minerals in particular that need to be added to a hepatic TF? Explain your answer.

14.   Calculate Mr. N's total energy needs using the Harris-Benedict formula and the appropriate stress factor.

15.   Calculate his protein needs.

16.   Mr. N is receiving D_5W at 75 cc/hr. How much fluid, grams of CHO, and kcals is this per day? Show your work.

17.   Considering the above, of the available hepatic formulas, which one would you choose and why?

18.   How much of this formula will Mr. N need to meet the requirements you previously calculated? Consider the I.V. kcals from D_5W.

19.   Give the starting strength and flow rate you would use and the progression to the final strength and flow rate. Explain your rationale.

20.   Considering that Mr. N is receiving lactulose and neomycin, can the TF be properly absorbed? Explain your answer.

Assume that Mr. N is going to recover from his hepatic encephalopathy; the edema and ascites will diminish significantly, and his renal function will return to normal. He will again be able to resume his diet. Also, assume that Mr. N will be discharged with end stage cirrhosis and will not be able to drink alcohol at all. He will still have some edema and some ascites, his serum albumin will be very low and he will be very weak.

21.   Assuming that Mr. N will go home on a diet, estimate what that diet will be. Include in your estimate the number of kcals he should have, the grams and percent of protein, the grams and percent of carbohydrate, the grams and percent of fat, the cc of fluid, the grams of Na, and any other restrictions or supplements that you feel are necessary.

Related References

1. Addolorato, G., Capristo, E., Stefanini, G.F., & Gasbarrini, G. (1998). Metabolic features and nutritional status in chronic alcoholics. *Am. J. Gastroenterol.* 93(4):665-6.

2. Cabre, E. & Gassull, M.A. (1994). Nutritional therapy in liver disease. *Acta. Gastroenterol. Belg.* 57(1):1-12.

3. Campillo, B., Bories, P.N., Pornin, B., & Devanlay, M. (1997). Influence of liver failure, ascites, and energy expenditure on the response to oral nutrition in alcoholic liver cirrhosis. *Nutrition.* 13(7-8):613-21.

4. Cerra, F.B., Cheung, N.K., Fischer, J.E., Kaplowitz, N., Schiff, E.R., Dienstag., J.L., Bower, R.H., Mabry, C.D., Leevy, C.M., & Kiernan, T. (1985). Disease-specific amino acid infusion (F080) in hepatic encephalopathy: a prospective, randomized, double-blind, controlled trial. *JPEN J Parenter Enteral Nutr.* 9(3):288-295.

5. Charlton, M.R. (1996). Branched chains revisited. *Gastroenterology.* 111(1):252-5.

6. de Ledinghen, V., Beau, P., Mannant, P.R., Borderie, C., Ripault, M.P., Silvain, C., & Beauchant, M. (1997). Early feeding or enteral nutrition in patients with cirrhosis after bleeding from esophageal varices? A randomized controlled study. *Dig Dis Sci.* 42(3): 536-41.

7. Fischbach, F.T. (1995). *A Manual of Laboratory & Diagnostic Tests.* 5th Ed. Philadelphia. J.B. Lippincott Company.

8. Gloria, L., Cravo, M., Camilo, M.E., Resende, M., Cardoso, J. N., Oliverira, A.G., Leitao, C.N., & Mira, F.C. Nutritional deficiencies in chronic alcoholics: relation to dietary intake and alcohol consumption. *Am. J. Gastroenterol.* 92(3):485-9.

9. Ichida, T., Shibasaki, K., Muto, Y., Satoh, S., Watanabe, A., Ichida, F. (1995). Clinical study of an enteral branched-chain amino acid solution in decompensated liver cirrhosis with hepatic encephalopathy. *Nutrition.* 11(2 Suppl):238-44.

10. Lieber, C.S., & Leo, M.A. (1998). Metabolism of ethanol and some associated adverse effects on the liver and stomach. *Recent Dev. Alcohol.* 14:7-40.

11. Marsano, L. & McClain, C.J. (1991). Nutrition and alcoholic liver disease. *J. Parenter. Enteral Nutr.* 15(3):337-44.

12. McEwen, D. (1998). End-stage alcoholism. *A.O.R.N. J.* 68(4):674-7.

13. Naylor, C.D., O'Rourke, K., Detsky, A.S., & Baker, J.P. (1989). Parenteral nutrition with branched-chain amino acids in hepatic encephalopathy. A meta-analysis. *Gastroenterology.* 97(4):1033-1042.

14. Nursing 98 Books. (1998). *Drug Handbook.* Springhouse, Pennsylvania. Springhouse Corporation.

15. Prendergast, A. & Fulton, F.L. (1997). *Medical Terminology: A Text/Workbook*. 4th Ed.

16. Pronsky, Z.M. & Solomon, E. (1998). *Food-Medication Interactions*. 10th Ed. Phoenix, Arizona. Food-Medication Interactions, Publishers and Distributors.

17. Tessari, P., Zanetti, M., Barazzoni, R., Biolo, G., Orlando, R., Vettore, M., Inchiostro, S., Perini, P., & Tiengo, A. (1996). Response of phenylalanine and leucine kinetics to branched chain-enriched amino acids and insulin in patients with cirrhosis. *Gastroenterology.* 111(1):127-37.

18. Whitney, E.N., Cataldo, C.B., & Rolfes, S.R. (1998). *Understanding Normal and Clinical Nutrition*, 5th Ed. West/Wadsworth.

CASE STUDY #34
ESOPHAGEAL CANCER AND CHEMOTHERAPY

INTRODUCTION
This is the first of two studies concerned with esophageal cancer. The location of the cancer adds to the nutritional problem since it affects swallowing. This case provides an introduction to the dietary treatment of chemotherapy patients as well as patients who have difficulty swallowing. The Interpretation of laboratory values and an understanding medical terminology are important in this case study.

SKILLS NEEDED

ABBREVIATIONS:
Knowledge of the following abbreviations is required in order to understand this case. You should learn these abbreviations before you begin to read the study.

Alk Phos : (ALT) alkaline phosphertase
ALT : alanine aminotransferase
AST : aspartate aminotransferase
BEE : basal energy expenditure
Bili : bilirubin
BMI : body mass index
BUN : blood urea nitrogen
CA : cancer
Cl : chloride
CPK : creatine phosphokinase
Cr : creatinine
CxR : chest x ray
Dx : diagnosis
FH : family history
g : gram
g/dl : gram per deciliter
Hct : hematocrit
Hgb : hemoglobin
K : potassium
kg : kilogram
LDH : lactic dehydrogenase

LES : lower esophageal sphincter
MCV : mean corpuscular volume
MD : medical doctor
mEq : milliequivalent
mg/dl : milligram per deciliter
mm/h : millimeter per hour
Na : sodium
NPO : nothing by mouth
qd : every day
RBW : reference body weight
Sed Rate : sedimentation rate
Ser Alb : serum albumin
TLC : total lymphocyte count
TPN : total parenteral nutrition
UBW : usual body weight
UGI : upper gastrointestinal
U/L : units per liter
WBC : white blood cell count
YOBF : year old black female
2° : secondary
μm^3 : cubic microns

LABORATORY VALUES:

You will need to be able to interpret the nutritional significance of the following laboratory values for this case study: Alk Phos, Bili, BUN, Cl, CPK, Cr, glucose, Hct, Hgb, K, LDH, lymphocytes, MCV, Na, Sed Rate, Ser Alb, AST, ALT, Platelet Count, Uric Acid, and WBC (Appendix B).

FORMULAS:

The formulas used in this case study include reference body weight, percent reference body weight, percent usual body weight, total lymphocyte count, BEE using the Harris-Benedict equation, stress factors, and total energy and protein needs, all of which can be found in Appendices A and D, Tables A - 7 through 10, 15 and D - 1 and 5.

MEDICATIONS:

Become familiar with the following medications before reading the case study. Note the diet-drug interactions, dosages and methods of administration, gastrointestinal tract reactions, etc.

1. Fudr (floxuridine).

Mrs. S is a 54 YOBF who is a buyer for one of the local department stores. She has lived a healthy life with no major illnesses. She is married and has three children, all of whom are married and well. Mrs. S has a FH of carcinoma. Her mother died of breast cancer at the age of 65 and her sister had a mastectomy for carcinoma when she was 62. Mrs. S has one younger brother who is healthy. She was doing fine until about 6 weeks before her first hospital admission, when she noticed some difficulty swallowing. She paid little attention to the problem at first but, as it seemed to get worse, she began to become concerned. This was a very busy time of the year for her. Christmas was approaching and sales were picking up. Mrs. S stayed in a continuous rush and had to "gulp" her food. She had heard people say that they "had a lump in their throat" when they were upset and decided, because of the stress and the fast eating, that was probably what she had. She tried eating more slowly and taking smaller bites, but it did not work. Her swallowing got worse. As a result, she was eating less and drinking more liquids. Mrs. S noted that she was losing weight but could not find time to go to the doctor.

The week after Christmas, six weeks after she first noticed a problem swallowing, she went to the doctor. She had lost 20 lbs since her last visit. He had some blood work done and did a UGI series on an outpatient basis. The results were as follows:

TEST	RESULT	NORM	TEST	RESULT	NORM	TEST	RESULT	NORM
Hgb	11.0 g/dl	12 - 16	LDH	1500 U/L	313 - 618	MCV	82.0 μm^3	82 - 98
Hct	35%	36 - 48	WBC	10.0x10³ /mm³	5 - 10 X 10³	Lymph	8.0%	20 - 40
Ser Alb	3.1 g/dl	3.9 - 5.0	Na	141 mEq/L	135 - 145	K	3.2 mEq/L	3.5 - 5.3
Glucose	82 mg/dl	65 - 110	Cl	104 mEq/L	98 - 106	Uric Acid	4.3 mg/dl	2.6 - 6.0
BUN	15 mg/dl	7 - 18	Cr	0.8 mg/dl	0.6 - 1.3	Sed Rate	32 mm/h	0 - 20
CPK	45 U/L	96 - 140	AST	13 U/L	5 - 40	ALT	8 U/L	7 - 56
Platelet Count	150,000/ mm³	140,000 - 400,000	Alk Phos	48 U/L	17 - 142	Bili	0.6 mg/dl	0.2 - 1.0

The UGI revealed an esophageal lesion. Mrs. S had to go back in two days for a CxR and an esophagogastroscopy. The endoscopy showed the lesion and a narrowing of the distal esophagus just above the LES. A biopsy was taken that proved to be positive for squamous cell carcinoma. The CxR was negative. Mrs. S was admitted to the hospital for more tests and surgery.

Mrs. S's Dx was:
▸ 1. Esophageal squamous cell CA
▸ 2. Dysphagia and odynophagia 2° to 1.
▸ 3. Microcytic, hypochromic anemia
▸ 4. Malnutrition

Her height was 5' 6" and she weighed 155 lbs. Her UBW was 175 lbs.
****

QUESTIONS:

1.   Determine Mrs. S's RBW, percent RBW, and percent UBW. Please show all calculations.

$$RBW = \frac{Actual}{Reference} \quad \frac{155}{136}$$

$$\%UBW = \frac{Actual}{Usual} = \frac{155}{175} \times 100$$

$$\%RBW = \frac{155}{136} \times 100$$

2.   Calculate Mrs. S's TLC.

$$White\ BC \times \%\ lymph = TLC$$

$$10 \times 10^3 \times 8.0 = 80 \times 10^4$$

3.   Which of the abnormal lab values could be indicative of cancer?

low hemoglobin   high LDH
low hematocrin
low serum albumin

4.   Which of the abnormal lab values would be indicative of malnutrition in general?

low hematocrin   low MCV
low serum albumin

5.   Which of the abnormal lab values would be indicative of anemia?

low Red Blood cells
low hemoglobin count
low hematocrin value

6.   Which of the abnormal lab values would be indicative of microcytic anemia?

low hemoglobin values
low hematocrin value

7.   Briefly define the following:

Mastectomy: *surgecol removal of breast*

Carcinoma: *malignant neoplasms*

Esophageal lesion: *abnormality of the lining of the essophagus*

Endoscopy: *a procedure used to view the esophagus, using a flexible tube passed into the stomach*

Esophagogastroscopy: *specifically looking into stomach & esophagus with endoscope*

Squamous cell carcinoma: *cancer of the lining cells of GI tract (superficial)*

Microcytic: *characterized by smaller than normal erythrocytes and less circulating hemoglobin (iron deficieney & thalassemia)*

Hypochromic: *characterized by deficient hemoglobin content of red blood cells*

Radiographic studies: *x-rays*

Dysphagia: *difficulty swallowing*

198 Odynophagia: *Pain, related to infections, neoplasms, or mechanical obstructions. Painful swallowing*

****

Additional radiographic studies suggested that the cancer was not as extensive as the MD originally thought, but Mrs. S was distraught just the same. She was extremely fearful of cancer because of her family history. She had had no idea that she was in such poor shape nutritionally. Mrs. S was so upset that she was not ready for surgery. She asked if she could have a few days to think about it. Considering her poor condition, her physician agreed and allowed the RD to work with her to improve her nutritional status. Meanwhile, he also decided to initiate chemotherapy in an attempt to shrink the tumor prior to surgery. His orders included:

quest 12

1. High protein, soft diet as tolerated.
2. At least three cans of high protein liquid supplement between meals qd.
3. floxuridine (Fudr) 0.5 mg/kg intraarterial qd.

The hospital was short staffed and was forced to use entry level RDs who did not have the usual orientation and training that the hospital provides before assigning someone to a unit. The RD assigned to Mrs. S had just graduated and had had very little experience with cancer patients. Try to picture the situation the RD is going into, something you should do before entering every hospital room. She is going to talk to a woman who has seen cancer kill several in her family and who went from being very healthy to very sickly in a short period of time. The patient is very upset.

The RD knocked on the patient's door and gained entry appropriately. She properly introduced herself and made sure that she was talking to the right patient. During the introduction, Mrs. S made it clear that she was very upset because she had just been told by her physician that she required surgery for cancer. The RD began her interview:

✗ **RD:** Mrs. S, I understand you are upset right now, but I need some information about your diet so that we can send you the foods you like to eat. ✗

Mrs. S: Honey, I don't care about eating anything right now, I'm so upset. Everything just sits in my stomach. It's so hard to swallow. It seems like it is getting harder every day, and it burns so.

RD: Well, have you noticed any particular foods that bother you more than others?

✗ **Mrs. S:** They all bother me. I can only swallow liquids. I just can't. . .

RD: How about blenderized or pureed foods? Have you tried any of them?

Mrs. S: No, but I have tried some mashed potatoes and I still was not . . . ✗

RD: Have you tried diluting the potatoes with milk?

Mrs. S: No, but I don't think that will work.

RD: Well you really don't know if it will work until you have tried it. How about if we send you some very dilute creamed potatoes and some pureed foods. We have very nice pureed food here.

Mrs. S: Honey, nobody has nice pureed food. I hate that stuff!

RD: Well, I want you to try ours before you pass judgment.

Mrs. S: I can't tolerate anything with acid. It burns so much. And I can't take anything that is highly seasoned.

RD: What about for breakfast? Do you eat eggs?

Mrs. S: No. Too high in cholesterol.

RD: Yes, but they are high in protein and very digestible. While you are sick we are not going to worry about cholesterol . . .

✗ **Mrs. S:** The doctor tells me not to eat it; you tell me to eat it. ✗

RD: What about juice?

Mrs. S: It burns, darling, it burns.

RD: Even grape juice?

Mrs. S: Even grape juice. They all burn. Honey, I don't want to be rude, but I don't feel like talking about food right now. I'm too upset. Just thinking about food makes me nauseous. Please come back tomorrow. Maybe I'll feel better.

RD: Well, OK. I'll let you off this time, but I'll be back. In the meantime, I'm going to send you some pureed food and some liquid supplements between meals. What is your favorite flavor?

Mrs. S: Strawberry.

RD: All right. I'm going to send you some strawberry supplements. You try to drink all of it and be sure to try to eat all the meat, and drink your milk. OK?

Mrs. S: Yea. OK.

****

QUESTIONS CONTINUED:

8. Define intraarterial. What dangers could accompany this method of administration?

   Putting something in the arteries (IV). Infections, bleeding, pain, spasum of artery (cold fingers)

9. List the mistakes the RD made during her conversation with Mrs. S.

   The way she started of by saying I understand..., because in actuality she does not understand, she should be more empathetic than sympathetic. When Mrs. S say the docter tells me this and you tell me that

10. List the segments of the interview you thought were well done.

    the RD came up with no response to that phrase. She should have explained that in her posistion she teaches patients about food, and knows about the interactions,..... I would do the interview entirely different, becoming

11. What would you do differently and how?

    more personal, asking her, and not telling her, It sounds like this RD is trying to force feed her. She also interups her patients they could be giving valuable information that the RD needs. Finding out what foods she likes

12. What energy and protein supplement would you use in this case?

    Vit.C inhibits the conversion of nitrates to nitroamnes. Derivatives of Vit. A ,antioxiant foods additives BHT ,and some materials in the cabbage family are thought to be anticarcinogens

13. Using the Harris-Benedict equation, determine Mrs. S's BEE. Using the appropriate stress factor, calculate her total energy needs (Appendix D, Tables D - 1 and 5).

$$655 + (9.6 \times wt) + (1.7 \times ht) - (4.7 \times age) = BEE$$

$$Energy = 1.5 \times BEE$$

(kg) (cm)

14. Determine her protein needs (Appendix D, Table D - 5).

1.5-2.0 g IBW kg

15. Considering all of the information above, outline a day's menu for Mrs. S. p 588

① Decrease fat intake to 30% of kilocalories.

② Increase intake of fruits, vegetables, and whole grains

③ minimize intake of cured, pickled, and smoked foods

④ minimize contamination of foods with carcinogens

⑤ If alcoholic beverages are consumed, intake should be moderate.

2,000 calories per day   High protein foods   cold foods
easy to swallow   oatmeal   baby foods ??
low acidity foods   texture of food   apricot + pear nectars
                                        are soothing

16. Look up the drug floxuridine and list all of the complications that pertain to nutritional status. given intraarterially, anticarcinogen.

nausea & diarreah, dehydration, vomiting

****

In the week that followed, Mrs. S developed just about every adverse reaction listed for floxuridine, including nausea and diarrhea. She ate very poorly but was able to drink the liquid supplements fairly well. After 10 days of treatment, everything either had a metallic taste or tasted too sweet. The smell of food made her nauseous. She was afraid to drink any more of the supplement because of the diarrhea.

****

QUESTIONS CONTINUED:

17.   Is there a different supplement that could be used in these circumstances? If so, identify the supplement and give the scientific basis for using it.

18.   There are a few things that can be done to help cancer patients overcome the metallic taste and nausea that occurs with smelling food. Be creative and see if you can think of anything that may be helpful in this respect.

Periodic use of an artificial saliva solution is also helpful, as is the frequent consumption of fluids to prevent dry mouth. If oral feeding is possible after surgery, general dietary recommendations would include liquid or soft textured, moist foods for easy mastication and swallowing, and small frequent meals are high in cal

19.   Sketch the location of Mrs. S's lesion.

Diaphragm

****

Mrs. S was still losing weight and her lab values were getting worse. She was becoming more depressed. The physician decided to make her NPO and initiated TPN. This was continued for one week before surgery and will be discussed in the next case study.

Related References

1. Burt, M.E., Gorschboth, C.M., & Brennan, M.F. (1982). A controlled, prospective, randomized trail evaluating the metabolic effect of enteral and parenteral nutrition in the cancer patient. *Cancer.* 15;49(6):1092-105.

2. Fischbach, F.T. (1995). *A Manual of Laboratory & Diagnostic Tests.* 5th Ed. Philadelphia. J.B. Lippincott Company.

3. Girvin, G.W., Matsumoto, G.H., Bates, D.M., Garcia, J.M., Clyde, J.C., & Lin, P.H. (1995). Treating esophageal cancer with a combination of chemotherapy, radiation, and excision. *Am. J. Surg.* 169(5):557-9.

4. Lee, J.H., Machtay, M., Unger, L.D., Weinstein, G.S., Weber, R.S., Chalian, A.A., & Rosenthal, D.I. (1998). Prophylactic gastrostomy tubes in patients undergoing intensive irradiation for cancer of the head and neck. *Arch. Otolaryngol. Head Neck Surg.* 124(8): 871-5.

5. Machin, J., & Shaw, C., (1998). A multidisciplinary approach to head and neck cancer. *Eur. J. Cancer Care.* 7(2):93-6.

6. Nelson, G.M. (1998). Biology of taste buds and the clinical problem of taste loss. *Anat. Rec.* 253(3):70-8.

7. Nursing 98 Books. (1998). *Drug Handbook.* Springhouse, Pennsylvania. Springhouse Corporation.

8. Pearlstone, D.B., Lee, J.I., Alexander, R.H., Chang, T.H., Brennan, M.F., & Burt, M. (1995). Effect of enteral and parenteral nutrition on amino acid levels in cancer patients. *J. Parenter. Enteral. Nutr.* 19(3):204-8.

9. Prendergast, A. & Fulton, F.L. (1997). *Medical Terminology: A Text/Workbook.* 4th Ed.

10. Pronsky, Z.M. & Solomon, E. (1998). *Food-Medication Interactions.* 10th Ed. Phoenix, Arizona. Food-Medication Interactions, Publishers and Distributors.

11. Righi, P.D., Reddy, D.K., Weisberger, E.C., Johnson, M.S., Trerotola, S.O., Radpour, S., Johnson, P.E., & Stevens, C.E. (1998). Radiologic percutaneous gastrostomy: results in 56 patients with head and neck cancer. *Laryngoscope.* 108(7):1020-4.

12. Schlag, P.M. (1992). Randomized trial of preoperative chemotherapy for squamous cell cancer of the esophagus. The Chirurgische Arbeitsgemeinschaft Fuer Onkologie der Deutschen Gesellschaft Fuer Chirurgie Study Group. *Arch. Surg.* 127(12):1446-50.

13. Sonis, S. & Clark, J. (1991). Prevention and management of oral mucositis induced by antineoplastic therapy. *Oncology.* 5(12):11-8, 18-22.

14. The Veterans Affairs Total Parenteral Nutrition Cooperative Study Group. (1991). Perioperative total parenteral nutrition in surgical patients. *N. Engl. J. Med.* 325(8):515-32.

15. Urba, S. (1997). Combined-modality treatment of esophageal cancer. *Oncology.* 11(Suppl 9):63-7.

16. Weisburger, J.H. (1998). Can cancer risks be altered by changing nutritional traditions? *Cancer.* 83(7):1278-81.

17. Whitney, E.N., Cataldo, C.B., & Rolfes, S.R. (1998). *Understanding Normal and Clinical Nutrition*, 5th Ed. West/Wadsworth.

18. Wilkes, J.D. (1998). Prevention and treatment of oral mucositis following cancer chemotherapy. *Semin. Oncol.* 25(5):538-51.

CASE STUDY #35
ESOPHAGEAL CANCER AND TPN

INTRODUCTION

This study continues with esophageal cancer but is for the advanced student. The case involves TPN formulation and monitoring.

SKILLS NEEDED

ABBREVIATIONS:

Knowledge of the following abbreviations is required in order to understand this case. You should learn these abbreviations before you begin to read the study.

BS : bowel sounds
cc/h : cubic centimeters per hour
CHO : carbohydrate
D/C : discontinue
D_5NS : 5% dextrose in normal saline
$D_{10}W$: 10% dextrose in water
Hct : hematocrit
Hgb : hemoglobin
I.V. : intravenous
MCV : mean corpuscular volume

MS : morphine sulfate
N/G : nasogastric
NPO : nothing by mouth
OR : operating room
po : by mouth
post-op : post operative
prn : as needed
Ser Alb : serum albumin
TF : tube feeding
TLC : total lymphocyte count
TPN : total parenteral nutrition

LABORATORY VALUES:

You will need to be able to interpret the nutritional significance of the following laboratory values for this case study: Hct, Hgb, lymphocytes, Ser Alb, and TLC (Appendix B).

FORMULAS:

The formulas used in this case study include simple percentages and the calculation of Kilocalorie and protein needs to determine TPN formulation.

MEDICATIONS:

Become familiar with the following medications before reading the case study. Note the diet-drug interactions, dosages and methods of administration, gastrointestinal tract reactions, etc.

1. Fudr (floxuridine); 2. Lomotil (diphenoxylate hydrochloride); 3. Morphine Sulphate

Ten days after Mrs. S's admission, her weight was 148 lbs (down 7 lbs). Diarrhea was a problem and she was not eating well. The adverse effects of floxuridine combined with her poor intake caused an additional decrease in Hgb, Hct, MCV, TLC, Ser Alb, and platelets. Her MD wrote the following orders:

- ▸ 1. D/C floxuridine
- ▸ 2. D/C diet and make pt NPO
- ▸ 3. 2 units of packed cells now
- ▸ 4. Have RD calculate TPN formula and flow rate
- ▸ 5. Schedule surgery in 7 days
- ▸ 6. Diphenoxylate hydrochloride (Lomotil)
- ▸ 7. D_5NS at 50 cc/h

The RD asked the support team dietitian to help her formulate the TPN. In order to do this, it is necessary to know the kcals in one gram of I.V. glucose.

****

QUESTIONS:

1.   How many kcals are provided daily by the D_5NS at 50 cc/h?

2.   Calculate a TPN formula for Mrs. S. In your calculations include the following:
         a. Calculate the grams of CHO and protein you would give per liter per day.
         b. List any additives you would use.
         c. Explain your flow rate progression from initial flow rate to final flow rate.
         d. Would you use IV lipids? Tell why or why not.
         e. If you used lipids, what percent would you use, how much would you use, and how
            would you administer (method of administration and flow rate)?
         f. How many total kcals and how many grams of protein is this per day?
         g. What lab values would you monitor and how often should they be checked?
         h. Discuss the use of prealbumin or retinol binding protein in this case.

3.   What is the mechanism of action of diphenoxylate hydrochloride (Lomotil)?

4.   What nutritional complications could it have?

****

Mrs. S tolerated the TPN well. The side effects of floxuridine disappeared and Mrs. S felt a lot better. By the time she went to the OR she felt stronger and was in much better spirits. Her surgery was uneventful and she recovered without a problem. A resection of the lower esophagus was done. The surrounding lymph nodes checked were negative for metastasis and the surgeons felt confident that they had removed all the cancer. A gastrostomy was also completed in surgery. The TPN was held during the surgery and $D_{10}W$ was hung in its place. TPN was restarted after Mrs. S's recovery.

Post-op orders included:

- ‣ 1. N/G to low intermittent suction
- ‣ 2. Continue NPO
- ‣ 3. Restart TPN at rate RD recommends
- ‣ 4. Have RD evaluate pt for proper TF and administration when BS return
- ‣ 5. MS for pain prn

****

QUESTIONS CONTINUED:

5.   At what flow rate and strength would you restart the TPN and how would you progress the flow rate?

6.   Discuss the function of the N/G tube to low intermittent suction.

7. Discuss the relationship between surgery, BS, and refeeding. Include in your discussion the effects of hypoalbuminemia and the appropriate tube feeding for this condition.

8. Research the mechanism of action of MS. Can you find any interaction this drug could have with the above relationship of BS and refeeding you just discussed?

9. Discuss why you would want to start the tube feeding as soon as possible.

10. Why was TPN discontinued and $D_{10}W$ hung in its place during surgery?

11. Discuss the alternative of using a tube feeding here instead of initiating TPN. In your discussion include the advantages and disadvantages of a tube feeding, what tube feeding you would use and why, and how you would recommend the feeding tube be placed.

12.   Describe the gradual increase of TF with the accompanying decrease of TPN and tell how long you would expect to take before meeting Mrs. S's requirements with TF alone.

13.   How long would it be before Mrs. S would start to eat again and what progression from TF to po feedings would you recommend?

14.   What final diet would you recommend for Mrs. S to go home with?

15.   Briefly define the following terms:

Resection:

Metastasis:

Gastrostomy:

▸ **ADDITIONAL OPTIONAL QUESTION** ◂

Tube Feeding Drill:

16. Using the table below, compare several of the enteral nutritional supplements that would be appropriate for Mrs. S. (☞ *Hint: Remember that everything tastes metallic or sweet and that she wants to swallow as little as possible.*)

Product	Producer	Form	Cal/ml	Non-pro Cal/g N	g/L			Na mg	K mg	mOs m/kg Water	Vol to meet RDAs in ml	g of fiber /L	Free water /L in ml
					Pro	CHO	Fat						

Related References

1.   Burt, M.E., Gorschboth, C.M., & Brennan, M.F. (1982). A controlled, prospective, randomized trail evaluating the metabolic effect of enteral and parenteral nutrition in the cancer patient. *Cancer.* 15;49(6):1092-105.

2.   Daly, J.M., Massar., E., Giacco, G., Fraazier, O.H., Mountain, C.F., Dudrick, S.J., & Copeland, E.M. 3d. (1982). Parenteral nutrition in esophageal cancer patients. *Ann Surg.* 196(2):203-8.

3.   Fischbach, F.T. (1995). *A Manual of Laboratory & Diagnostic Tests.* 5th Ed. Philadelphia. J.B. Lippincott Company.

4.   Girvin, G.W., Matsumoto, G.H., Bates, D.M., Garcia, J.M., Clyde, J.C., & Lin, P.H. (1995). Treating esophageal cancer with a combination of chemotherapy, radiation, and excision. *Am. J. Surg.* 169(5):557-9.

5.   Meguid, M.M. & Meguid, V. Preoperative indentification of the surgical cancer patient in need of postoperative supportive total parenteral nutrition. *Cancer.* 1:55(Suppl):258-62.

6.   Nursing 98 Books. (1998). *Drug Handbook.* Springhouse, Pennsylvania. Springhouse Corporation.

7.   Pearlstone, D.B., Lee, J.I., Alexander, R.H., Chang, T.H., Brennan, M.F., & Burt, M. (1995). Effect of enteral and parenteral nutrition on amino acid levels in cancer patients. *J. Parenter. Enteral. Nutr.* 19(3):204-8.

8.   Prasad, A.S, Beck., F.W., Doerr, T.D., Shamsa, F.H., Penny, H.S., Marks, S.C., Kaplan, J., Kucuk, O., & Matog, R.H. (1998). Nutritional and zinc status of head and neck cancer patients: an interpretive reivew. *J. Am. Coll. Nutr.* 17(5);409-18.

9.   Prendergast, A. & Fulton, F.L. (1997). *Medical Terminology: A Text/Workbook.* 4th Ed.

10.   Pronsky, Z.M. & Solomon, E. (1998). *Food-Medication Interactions.* 10th Ed. Phoenix, Arizona. Food-Medication Interactions, Publishers and Distributors.

11.   Sikora, S.S., Ribeiro, U., Kane III, J.M., Landreneau, R.J., Lembersky, B., & Posner, M.C., Role of nutrition support during induction chemoradiation therapy in esophageal cancer. *J. Parenter. Enteral Nutr.* 22(1):18-21.

12.   The Veterans Affairs Total Parenteral Nutrition Cooperative Study Group. (1991). Perioperative total parenteral nutrition in surgical patients. *N. Engl. J. Med.* 325(8):515-32.

13.   Whitney, E.N., Cataldo, C.B., & Rolfes, S.R. (1998). *Understanding Normal and Clinical Nutrition*, 5th Ed. West/Wadsworth.

CASE STUDY #36
AIDS

INTRODUCTION
A patient with AIDS could have pneumonia, a liver complication, renal involvement, endocarditis, or any combination of the above. Usually with these problems, diarrhea, GI cramps, and infections of the mouth and esophagus are common. This complicates nutritional therapy and frequently a TF or TPN is necessary. Each case is different and has to be evaluated with the general nutritional goals plus the additional goals for each complication.

SKILLS NEEDED

ABBREVIATIONS:
Knowledge of the following abbreviations is required in order to understand this case. You should learn these abbreviations before you begin to read the study.

AIDS : acquired immune deficiency syndrome
ARC : AIDS related complex
cc/hr : cubic centimeters per hour
CHO : carbohydrate
D_5W : 5% dextrose in water
F : Fahrenheit
g/dl : grams per deciliter
GI : gastrointestinal
HIV : human immunodeficiency virus

I.V. : intravenous
N/V : nausea and vomiting
RBW : reference body weight
Ser Alb : serum albumin
TF : tube feeding
TPN : total parenteral nutrition
WBC : white blood cell count
YOWM : year old white male
@ : at
¼NS : one-fourth strength normal saline

FORMULAS:
The formulas used in this case study include: reference body weight, caloric needs using the Harris-Benedict equation, total caloric and protein requirements, simple percentages for tube feeding and I.V. calculations, and flow rates (Appendices A and D, Tables A - 7 through 10 and D - 1 and 5).

MEDICATIONS:
Become familiar with the following medications before reading the case study. Note the diet-drug interactions, dosages and methods of administration, gastrointestinal tract reactions, etc.

1. Retrovir (zidovudine, or formally AZT); 2. Vincasar PFS (vincristine sulfate); 3. Mycostatin (nystatin).

LT is a 35 YOWM who is employed in a major department store. He is a buyer for the men's clothing department and has been successful in his job for the past 10 years. He brags about how hard he works during the week and how hard he parties on weekends. He calls himself a fast-lane single and does not hesitate to admit that he has tried various drugs. As with many drug users, the pills and marijuana led to I.V. drug abuse. About 2 years ago, he started to notice that he was becoming weaker and easily fatigued. His appetite was poor and he was losing weight. He woke up during the night soaking wet with severe sweats. Before the symptoms started, LT weighed 190 lbs and was 6'1". When he was approaching 170 lbs, he went to a physician. Some blood work was done and indicated that he had elevated WBC, elevated sedimentation rate, and a normocytic normochromic anemia. He was admitted to the hospital. Further tests indicated that he had infective endocarditis. He underwent antibiotic therapy and was sick for several weeks. He continued to lose weight to 165 lbs. While in the hospital, he was also tested for AIDS since endocarditis is not an uncommon complication of I.V. drug abuse. LT tested HIV+. LT did not show any symptoms of AIDS in this hospitalization, but his physician told him that he should be evaluated frequently. Any change in his health status should be reported to the physician immediately.

Since LT had been using I.V. drugs for some time, it was not known how long LT had been infected with the AIDS virus. Not long after LT's discharge from the hospital, he began to have generalized lymphadenopathy. He returned to his physician and was diagnosed as having ARC. During this time his temperature was elevated and he was still losing weight. He was beginning to experience some abdominal cramps and diarrhea. His physician suggested that he start on zidovudine (Retrovir). Not long after this, LT noticed an unusual bump on his chest. When he went back to his physician, it was diagnosed as Kaposi's sarcoma. He was given vincristine sulfate (Vincasar PFS) to treat the Kaposi's sarcoma.

****

QUESTIONS:

1.   List the symptoms of AIDS that were manifested in LT.

① weaker & easily fatigued .   ④ generalized lymphadeno pathy
② Poor appetite & losing weight
③ elevated WBC

2.   What are the general nutritional goals for a patient with AIDS?

Achieving nutritional health and preventing malnutrition are essential in maintaining positive health occurances for persons with HIV, and medical nutrition therapy in HIV care must be included as a primary component of total health care. ① Screening, referral, assessment, intervention, outcomes evaluation & communication

3.   Calculate LT's total energy needs using the Harris-Benedict equation and appropriate stress factor (Appendix D, Tables D - 1 and D - 5).

66 + (13.7 × wt) + (5 × ht) - (6.8 × age)

66 + (13.7 × 75) + (5 × 15.49) - (6.8 × 35)

66 + 1027.5   +   77.45   -   238

932.95

Use of doxorbin hydrelonde, for example, requires adequate riboflavin to reduce drug toxicity. Excellent nutritional status enhances the ability to fight subsequent infections.

4.   Calculate his protein needs.

$$932.95 \times .3 = 279.885$$

5.   Are there any vitamins or minerals LT should be taking in therapeutic doses? Explain your answer. *Studies suggest the need for increased intake of the following micronutrients: beta carotene, vitamin E, ascorbic acid, vitamin B12, vitamin B6 and folic acid. Persons consuming an inadequate diet use a vitamin-mineral supplement providing 100% of the RDA's*

6.   Give the mechanism of action of the following drugs:

Vincristine sulfate (Vincasar PFS): *insure adequate fluid intake* *antineoplastic, Vinca Alkaloid*

Zidovudine (Retrovir): *Antiretroviral*

7.   List any nutritional complications of each of these drugs:

Vincristine sulfate (Vincasar PFS): *Anorexia, loss of weight*

Zidovudine (Retrovir): *hyper o hypotension*

****

Not long after he had taken these medications, LT's symptoms increased with an elevated temperature (which was now 100.6° F), N/V, headaches, diarrhea, malaise, and extreme fatigue. He had to take a leave of absence from work. He developed a bad cough and chest congestion. His physician admitted him to the hospital for tests and found LT to have a Pneumocystis pneumonia caused by Pneumocystis carinii. During LT's hospital stay, he continued to have complications. He experienced infections such as candidiasis. This made it very difficult for him to eat and required another medication, nystatin (Mycostatin). LT also experienced severe diarrhea and he became very depressed. His Ser Alb was 2.2 g/dl. His desire to eat was affected, so the RD recommended that LT be given a tube feeding. The physician agreed. A feeding tube was placed into the small bowel and two I.V.s were infusing, one with ¼NS @ 25 cc/hr for medications and one with D_5W @ 50 cc/hr.

****

QUESTIONS CONTINUED:

8.   What diet therapy would you recommend for LT considering the above conditions? Would his stress factor and energy needs change? Explain your answer.

9.   Why would candidiasis make it difficult to eat? *Common in persons with AIDs. Symptoms include soreness of the mouth and tongue often described as a burnt feeling and pain or difficulty swallowing.*

10.   What is nystatin (Mycostatin)? List its possible adverse reactions.

anti candidiasis, anti fungal. Can cause GI distress Nausea & vommiting, stomach pain, diareah, no GI absorption. Dental care must be done carefully.

11.   What is the significance of the Ser Alb being 2.2 g/dl? Discuss the pathophysiological effects this could have on LT's GI tract.

12.   From LT's history, summarize all of the possible causes of diarrhea and explain how they might cause diarrhea.

13.   Discuss what could be done, if anything, to check LT's diarrhea.

14.   What tube feeding would you recommend for LT? Explain your choice.

15.   How many grams of CHO and kcals is LT receiving from the D_5W @ 50 cc/hr? Show your work.

16.   How would you manage the TF, i.e., what strength and flow rate would you start with and how would you progress to the final rate? Include the final rate that you hope to achieve.

17.   Assuming you reach the final rate in your goals, how many kcals and how much protein will this provide? Show your work.

18.   Add to this the kcals from the D_5W and compare the total to the energy requirement you previously calculated. Discuss your results.

19.   What criteria would you use to suggest a change from enteral to parenteral nutrition?

20.   Define the following terms:

Lymphadenopathy: swollen, firm and sometimes tender lymphnodes secondary to any number of causes ranging from infection.

Normocytic Normochromic Anemia: normal small hemoglob in content.

889 Kaposi's sarcoma: a malignant neoplastic vascular proliferation characterized by the development of bluish-red cutaneous nodules

→ Endocarditis: infection on a heart valve

HIV Virus: Acute Human immunodeficiency virus infection an acute syndrome characterized by fever, malaise lymphadenopathy......

— Pneumocystis pneumonia: bacterial infection, common pneumonia infection in HIV patients

— Pneumocystis carinii: the organism that causes the infection

— Malaise: fatigue or tiredness

Candidiasis: an infection caused by the yeast like fungus candida.

463 Enteral: provisions of nutrients to the gastro intestinal tract through a tube or catheter when oral intake in inadequate

463 Parenteral: provisions of nutrients through the blood stream intravenously

— ARC: Aids related complex some of the manifestation of AIDs

▸ ADDITIONAL OPTIONAL QUESTION ◂

Tube Feeding Drill:

21.   Some research indicates that the addition of glutamine, arginine, and/or fish oils may be of some benefit to immunocompromised patients, patients with atrophic GI tracts, and/or edematous guts. Using the table below, compare several of the enteral nutritional supplements that would be appropriate for one or more of these conditions.

Product	Producer	Form	Cal/ml	Non-pro Cal/g N	g/L			Na mg	K mg	mOs m/kg Water	Vol to meet RDAs in ml	g of fiber /L	Free water /L in ml
					Pro	CHO	Fat						

Related References

1. Beaugerie, L., Carbonnel, F., Carrat, F., Rached, A.A., Masio, C., Genre, J.P., Rozenbaum, W., & Cosnes, J. (1998). Factors of weight loss in patients with HIV and chronic diarrhea. *J. Acquir. Immune Defic. Syndr. Hum. Retrovirol.* 1;19(1);34-9.

2. Crotty, B., McDonald, J., Mijch, A.M., & Smallwood, R.A. (1998). Percutaneous endoscopic gastrostomy feeding in AIDS. *J. Gastroenterol. Hepatol.* 13(4):371-5.

3. Davidhizar, R. & Dunn, C. (1998). Nutrition and the client with AIDS. *J. Pract. Nurs.* 48(1):16-28.

4. Fischbach, F.T. (1995). *A Manual of Laboratory & Diagnostic Tests.* 5th Ed. Philadelphia. J.B. Lippincott Company.

5. Gerrior, J.L., Bell, S.J., & Wanke, C.A. (1997). Oral nutrition for the patient with HIV infectiion. *Nurs. Clin. North Am.* 32(4):813-30.

6. Kotler, D.P. (1998). Human immunodeficiency virus-related wasting: malabsorption syndromes. *Semin. Oncol.* 25(2 Suppl 6): 70-5.

7. Matarese, L.E. (1998). Enteral feeding solutions. *Gastrointest. Endosc. Clin. N. Am.* 8(3):593-609.

8. Moscardini, C., Touger-Decker, R., & Ostrowski, M.B. (1997). Nutritional needs in the AIDS patient. Recognizing and treating wasting syndrome. *Adv. Nurse Pract.* 5(6):34-7, 41-2.

9. Nursing 98 Books. (1998). *Drug Handbook.* Springhouse, Pennsylvania. Springhouse Corporation.

10. Pichard, C., Sudre, P., Karsegard, V., Yerly, S., Slosman, D.O., Delley, V., Perrin, L. & Hirschel, B. (1998). A randomized double-blind controlled study of 6 months or oral nutritional supplementation with arginine and omega-3 fatty acids in HIV-infected patients. Swiss HIV Cohort Study. *AIDS.* 1;12(1):53-63.

11. Prendergast, A. & Fulton, F.L. (1997). *Medical Terminology: A Text/Workbook.* 4th Ed.

12. Pronsky, Z.M. & Solomon, E. (1998). *Food-Medication Interactions.* 10th Ed. Phoenix, Arizona. Food-Medication Interactions, Publishers and Distributors.

13. Rabeneck, L., Palmer, A., Knowles, J.B., Seidehamel, R.J., Harris, C.L., Merkel, K.L., Risser, J.M., & Akrabawi, S.S. (1998). A randomized controlled trial evaluating nutrition counseling with or without oral supplementation in malnourished HIV-infected patients. *J. Am. Diet. Assoc.* 98(4):434-8.

14. Salomon, S.B., Jung, J., Voss, T., Suguitan, A., Rowe, W.B., & Madsen, D.C. (1998). An elemental diet containing medium-chain triglycerides and enzymatically hydrolyzed protein can improve gastrointestinal tolerance in people infected with HIV. *J Am Diet Assoc. 98(4):460-2.*

15.   U.S. National Library of Medicine. (1997). Important Therapeutic Information on Prevention of Recurrent Pneumocystis Carinii Pneumonia in Persons with AIDS. http://www.nlm.nih.gov/

16.   Whitney, E.N., Cataldo, C.B., & Rolfes, S.R. (1998). *Understanding Normal and Clinical Nutrition*, 5th Ed. West/Wadsworth.

CASE STUDY #37
HYPERTENSION

INTRODUCTION

This is the first of a three-part series about a man with a family history of hypertension and renal disease. The patient is not compliant with diet or medication orders. This is a basic study that concerns dietary sodium restriction and teaching techniques. As you read the RD's interview with the patient, pay attention to any inappropriate questions or mannerisms she uses and omissions she may make. If this patient had been compliant with diet and medications, the problems involved in the next two case studies may not have occurred. Review foods high in sodium and potassium.

SKILLS NEEDED

ABBREVIATIONS:

Knowledge of the following abbreviations is required in order to understand this case. You should learn these abbreviations before you begin to read the study.

BEE : basal energy expenditure
BMI : body mass index
BP : blood pressure
g : gram
HTN : hypertension
mg : milligram

Na : sodium
prn : as needed
qd : every day
qod : every other day
RBW : reference body weight
UBW : usual body weight
YOBM : year old black male

FORMULAS:

The formulas used in this case study include reference body weight, percent reference body weight, percent usual body weight, adjusted body weight, body mass index, caloric needs, and protein requirements (Appendices A and D, Tables A - 7 through 13 and D - 1 and 5).

MEDICATIONS:

Become familiar with the following medications before reading the case study. Note the diet-drug interactions, dosages and methods of administration, gastrointestinal tract reactions, etc.

1. Tenormin (atenolol); 2. Lasix (furosemide).

Mr. G is a 45 YOBM employed as a plant manager in the automobile industry. He enjoys his job, but it is highly stressful because of the pressure to produce as many cars as possible in a short period of time. Recently, the pressure has increased because of increased gas prices and decreased car sales. Mr. G is a worrier. He has been losing sleep over the slow economy and the effect it may have on his job. He is married and has children, one in high school and two in elementary school. His wife works as a cashier in a food store and sometimes has to work evenings. This places increased stress on Mr. G since he does not know how to cook and the children always complain when he tries.

He has a history of HTN but has not been checking it as he should. He had a routine physical recently and his BP was 175/100. This added to his concern because if it does not go down, he fears he will be forced by the plant physician to take sick leave. His medication included Tenormin, 50 mg qd and Lasix, 30 mg qod or prn for pedal edema. He has been on hypertensive medication for seven years but the strength of the medication has been increasing for the last three years because of the gradual increase in his BP. Mr. G has not been a compliant patient. He frequently does not take his medication because he feels fine and believes he is too young to have high BP, even though his father died from a stroke at 57. He has an uncle on his father's side who has renal failure secondary to untreated HTN.

MR. G is 5'10" and weighs 225 pounds, has a medium frame, and is moderately active. He has always been on the heavy side but his weight has increased by 30 lbs in the past year. At his last physical, the MD found more pedal edema than usual. Mr. G admitted to noncompliance in respect to his medication. The MD instructed him on the importance of taking his medication, talked to him about the effects of worrying, and told him to see the plant dietitian. It was the dietitian's job to complete a nutritional assessment and determine his energy and protein needs. She also had an order from the MD to instruct him on a 2 g Na diet with appropriate kcals and protein based on her assessment.
****

QUESTIONS:

1.   Determine Mr. G's RBW and percent of RBW (Appendix A, Table A - 7 through 9).

2.   Determine his percent of UBW (Appendix A, Table A - 7 through 9).

3.   Determine Mr. G's adjusted body weight (Appendix A, Table A - 11).

4.   Calculate his BMI (Appendix A, Table A - 12 and 13).

5. Determine his protein needs (Appendix D, Table D - 5).

6. Using the Harris-Benedict equation, calculate his BEE (Appendix D, Table D - 1).

7. Considering Mr. G as moderately active, calculate his total caloric needs for the day (Appendix D, Tables D - 3 and 5).

8. Mr. G's BP was 175/100. What does blood pressure mean? Is 175/100 something to be concerned about? Explain.

9. What are the functions of Tenormin and Lasix?

10. List any nutritional complications of Tenormin and Lasix.

11.   List the principles of a 2 g Na diet.

****

The RD had Mr. G come to his office for a discussion. He obtained a typical day's recall and a food frequency as follows:

RD: Mr. G, the plant MD tells me that you have been gaining weight recently and that your blood pressure has been elevated. Is that correct?

Mr. G: Yes, it is.

RD: OK. He wants me to talk to you about a diet. Have you ever been on any kind of diet?

Mr. G: Well, yea. I'm supposed to be watching my salt intake right now.

RD: How have you been doing?

Mr. G: Very good. I don't think I use that much.

RD: I see. What about a weight reduction diet? Have you ever been on a weight reduction diet?

Mr. G: Not really. I've tried to lose a few pounds every now and then, but nothing other than that.

RD: OK. When the doctor weighed you the other day, you were 225 lbs and 5'10". Is that correct?

Mr. G: That is what he said.

RD: OK. I want you to take me through a typical day from the time you get up in the morning, to the time you go to bed at night. Tell me everything you have to eat or drink; amounts, how it was prepared, everything. So, when you get up in the morning, what is the first thing you have to eat or drink?

Mr. G: Well, the first thing I have is a cup of coffee while I'm getting ready to go to work. Then I sit down and have some breakfast with my wife. I usually have a couple of sausage-biscuits with another cup of coffee. That's all I really eat in the morning. That's all I have time for.

RD: What is the next thing you have to eat or drink?

Mr. G: Well, I get to work and I start on my paperwork first. Then I have to make my rounds and check my men. I come back to the office and take a little break. I have a cup of coffee.

RD: OK. What is the next thing?

Mr. G: For lunch my wife fixes me a lunch box. I'll have a sandwich, piece of fruit, some cookies, a piece of cake if she made some the night before, you know, something like that. Some coffee.

RD: What kind of sandwich do you have?

Mr. G: Well, some kind of meat, you know, like luncheon meat, boiled ham, bologna, a piece of cheese on it, you know. The stuff you get in the store in the little packs.

RD: Does she put anything else on the sandwich?

Mr. G: Yea, sometimes she puts a little mustard or mayo.

RD: What kind of cheese?

Mr. G: I don't know, my wife gets that.

RD: What kind of fruit?

Mr. G: Usually an apple.

RD: What is the next thing you have to eat or drink?

Mr. G: Well, in the afternoon I usually get a soda.

RD: OK.

MR. G: Then when I go home I eat a good supper. I usually have some kind of meat. Could be meat loaf, chicken, some kind of meat. And we usually . . .

RD: Let me see if I can get you to accurately estimate how much meat you normally eat. These are plastic food models. Look at this model of roast beef and tell me if what you normally eat is smaller than, equal to, or larger than this piece.

Mr. G: Oh, it would be more than that. It would be about twice that.

RD: OK.

MR. G: Is that plastic food? Let me see that. Boy, that feels funny huh?

RD: What else?

Mr. G: Well, then we have some potatoes, you know . . .

RD: How are those potatoes fixed?

Mr. G: Sometimes she boils them, but we have french fries a lot. The kids like french fries.

RD: OK. What else would you have?

Mr. G: She always has some vegetables, like green beans, greens, or squash. Always some vegetables.

RD: How does she fix the vegetables?

Mr. G: Well . . . I don't know how she cooks them . . . what do you mean?

RD: Does she boil them? Fry them? Does she cook them with butter, margarine, bacon fat, or some other seasoning?

Mr. G: Every now and then we have some fried okra or squash, but other than that we don't have much fried foods. She does put lots of seasoning in the vegetables, like fat meat, bacon, stuff like that.

RD: OK. Mr. G, do you know if your wife adds any salt to your food?

Mr. G: Some, I don't know how much. You will have to ask her that.

RD: Does she ever try cooking with other seasonings and no salt?

Mr. G: I don't know. She does all the cooking.

RD: Mr. G, do you have any salad or bread with your meals?

Mr. G: We usually have some bread, some salad every now and then. Sometimes she fixes a pie or a cake.

RD: What do you put on your bread?

Mr. G: I put a little bit of margarine on it.

RD: What about the salad, do you put anything on your salad?

Mr. G: I use a little bit of salad dressing.

RD: What kind do you use, Mr. G?

Mr. G: Oh I don't know, she gets all kinds; whatever she has.

RD: OK. And what do you have to drink with your meal?

Mr. G: Kool Aid.

RD: Sweetened or unsweetened?

Mr. G: Sweetened.

RD: How much sugar does she add to the Kool Aid?

Mr. G: I don't know that. You will have to ask the wife.

RD: After supper is there anything you have to eat or drink before you go to bed?

Mr. G: Well, after we get the kitchen cleaned up, I like to help the kids with their homework. Then I like to relax, you know. I put my feet up and look at some TV. I might snack on some potato chips or peanuts and drink a beer or two before bed. See I really don't eat that much; I really don't know why I'm gaining all this weight.

RD: OK. Mr. G, you only mentioned fruit at lunch. Do you ever have fruit any other time?

Mr. G: No, that's about it. I don't care for fruit that much.

RD: Mr. G, I noticed that you did not mention milk. Do you drink milk?

Mr. G: No, I'm allergic to milk.

RD: You're allergic to milk? What happens when you drink milk?

Mr. G: Oh I get a lot of cramps and gas, you know.

RD: What about cheese or yogurt, do you eat much cheese?

Mr. G: Yogurt? I don't eat that stuff, but I like cheese. Cheese doesn't bother me. I eat some cheese and crackers sometimes at night with my beer.
****

QUESTIONS CONTINUED:

12. Identify any mistakes the RD may have made in this interview.

13. Is there anything you would do differently? Explain.

14. What are some good points about this interview?

15. What are the goals of nutritional therapy for Mr. G?

16.　In one column, list the foods that Mr. G eats that are high in sodium. In a second column, list possible substitutes for the foods in column one.

HIGH Na FOODS　　　　　　　　　　**SUGGESTED SUBSTITUTES**

17.　In column one, list the foods Mr. G eats that are high in fat. In a second column, list possible substitutes for the foods in column one.

HIGH FAT FOODS　　　　　　　　　**SUGGESTED SUBSTITUTES**

18.　Based on the recall, approximate Mr. G's energy and protein intake. Is he taking in too much, too little, or an appropriate amount?

19.　Approximate Mr. G's sodium intake (high, low, or appropriate).

20.　Explain the relationship between BP and pedal edema.

21.   Should the RD be concerned about Mr. G's calcium intake? Explain.

22.   Considering Mr. G's medication, diet, and recall, outline the teaching program you would use for him. Include in your outline behavioral changes you would recommend, foods to avoid with appropriate substitutes, and foods to include. Explain why you are making changes.

Related References

1.   Cook, N.R., Kumanyika, S.K., & Cutler, J.A. (1998). Effect of change in sodium excretion on change in blood pressure corrected for measurement error. The Trails of Hypertension Preventiion, Phase I. *Am. J. Epidemiol.* 1;148(5):431-44.

2.   Fischbach, F.T. (1995). *A Manual of Laboratory & Diagnostic Tests.* 5th Ed. Philadelphia. J.B. Lippincott Company.

3.   He, J., & Whelton, P.K. (1997). Role of sodium reduction in the treatment and prevention of hypertension. *Curr. Opin. Cardiol.* 12(2):202-7.

4.   Hunt, S.C., Cook, N.R., Oberman, A., Cutler, J.A., Hennekens, C.H., Allender, P.S., Walker, W.G., Whelton, P.K., & Williams, R.R. (1998). Angiotensinogen genotype, sodium reduction, weight loss, and prevention of hypertension: trials of hypertension prevention, phase II. *Hypertension.* 32(3):393-401.

5.   Katz, J. (1998). Salt wars. *Science.* 25;28(5385):1963.

6.   Kuller, L.H. Salt and blood pressure: population and individual perspectives. *Am. J. Hypertens.* 10(5 Pt 2):29S-36S.

7.   Luft, F.C., Morris, C.D., & Weinberger, M.H. (1997). Compliance to a low-salt diet. *Am. J. Clin Nutr.* 65(2 Suppl):698S-703S.

8.   McCarron, D.A. (1998). Diet and blood pressure--the paradigm shift. *Science.* 14;281 (5379):933-4.

9.   Nursing 98 Books. (1998). *Drug Handbook.* Springhouse, Pennsylvania. Springhouse Corporation.

10.   Prendergast, A. & Fulton, F.L. (1997). *Medical Terminology: A Text/Workbook.* 4th Ed.

11.   Pronsky, Z.M. & Solomon, E. (1998). *Food-Medication Interactions.* 10th Ed. Phoenix, Arizona. Food-Medication Interactions, Publishers and Distributors.

12.   Taubes, G. (1998). The (political) science of salt. *Science.* 14;281(5379):898-901,903-7.

13.   Weir, M.R., & Dworkin, L.D. (1998). Antihypertensive drugs, dietary salt, and renal protection: how low should you go and with which therapy? *Am J Kidney Dis.* 32(1):1-22.

14.   Whitney, E.N., Cataldo, C.B., & Rolfes, S.R. (1998). *Understanding Normal and Clinical Nutrition,* 5th Ed. West/Wadsworth.

CASE STUDY #38
RENAL DISEASE

INTRODUCTION

This case study is a continuation of #37. It demonstrates the possible effects of patients' noncompliance to their medication and diet regimen. It is a good introduction to renal disease, hemodialysis, and the associated diet, drugs, terms, and lab values.

SKILLS NEEDED

ABBREVIATIONS:

Knowledge of the following abbreviations is required in order to understand this case. You should learn these abbreviations before you begin to read the study.

A-V : arteriovenous
BEE : basal energy expenditure
BUN : blood urea nitrogen
Ca : calcium
c : with
cc : cubic centimeters
Cl : chloride
Cr : creatinine
d : day
D/C : discontinue
ESRD : end stage renal disease
g : gram
g/dl : grams per deciliter
Hct : hematocrit
Hgb : hemoglobin
HTN : hypertension
K : potassium
kg : kilogram
L : liter

MCV : mean corpuscular volume
mEq/L : milliequivalent per liter
Mg : magnesium
mg : milligram
mg/dl : milligram per deciliter

MNT : medical nutrition therapy
Na : sodium
neg : negative
P : phosphorous
pH : measure of acidity or alkalinity
p.o. : by mouth
prot : protein
p̄ : after
q A.M. : every AM or morning
qd : every day
RBW : reference body weight
S.C. : subcutaneous
Ser Alb : serum albumin
SG : specific gravity
t.i.d. : three times a day
u : units
3x/wk : three times per week
μm^3 : cubic microns
↑ : more or increases
φ : none or zero
T : one
viii : eight
x : ten

FORMULAS:

The formulas used in this case study include reference body weight, caloric needs using the Harris-Benedict equation, total caloric and protein requirements, and conversion of mEq to mg (Appendices A and D, Tables A - 5 and 7 and D - 1 and 5).

LABORATORY VALUES:
You will need to be able to interpret the nutritional significance of the following laboratory values for this case study: BUN, Ca, Cl, Cr, glucose, Hct, Hgb, K, MCV, Mg, Na, P, pH, uric acid, and Ser Alb.

MEDICATIONS:
Become familiar with the following medications before reading the case study. Note the diet-drug interactions, dosages and methods of administration, gastrointestinal tract reactions, etc.

1. Tenormin (atenolol); 2. Lasix (furosemide); 3. Phos-Lo (calcium acetate); 4. Epogen (epoetin alfa).

After talking to the dietitian, Mr. G followed his diet for a few months. During that time he lost approximately 18 lbs. He gradually felt better about himself but, as time passed, he went off his diet. At first, he took his medications as ordered, but then he started to skip them on occasion. He noticed that he had to go to the bathroom more frequently. He also noted that he had been waking up during the night with leg cramps, something he had never experienced in the past. He was not sure why this was happening but it started after he began taking the new medication. He believed the medication was causing the leg cramps, even though he did not know what the relationship was. He observed that if he took Lasix every other day instead of daily, he did not have as many leg cramps. Gradually he started taking Lasix less and less regularly. The pedal edema went away. He knew Lasix helped prevent edema, so he thought he didn't need to take it any longer. His condition remained stable for several months. When edema returned, he would take Lasix until it went away and then he would quit taking it. Every time he had a doctor's appointment he would take Lasix and his blood pressure medication several days before the appointment. He thought this would prevent obvious signs of edema and his blood pressure checks would be normal. It seemed to work. His blood pressure was borderline or slightly elevated. After the doctor's visit, he would quit taking his medication because he thought he did not need it.

This continued for a couple of years. Then he started to notice that when the edema came back and he took Lasix, it would not go away. He did not have to go to the bathroom as much. He also noticed some pain in the lower left flank. The pain persisted but he thought it was because he was not getting enough exercise. He did not relate the pain to renal disease, so he ignored it. The pain continued and the pedal edema got worse. He was gaining weight, but he really did not feel like he was eating any more than usual. He went off his diet by eating more fatty foods but continued to watch his sodium intake for almost a year. By then he was completely back to his old eating habits. One day he felt very fatigued and decided to go get his blood pressure checked by the plant nurse. It was 180/105. The nurse made an appointment for him to see his doctor. When the doctor weighed him, he had gained all of his weight back and then some. He now weighed 235 lbs. The doctor made him an appointment to see a nephrologist.

Mr. G went to see the nephrologist who examined him and had the following lab work done:

Urinalysis:

SG 1.001   Protein +2   Glucose neg.   Ketones neg.   pH 8.1

Table I

TEST	RESULT	NORM	TEST	RESULT	NORM	TEST	RESULT	NORM
Hgb	12.8 g/dl	14 - 17	MCV	77 µm³	82 - 98	Glucose	110 mg/dl	65 - 110
Hct	40%	42 - 52	Mg	2.1 mEq/L	1.3 - 2.1	Na	147 mEq/L	135 - 145
K	6.0 mEq/L	3.5 - 5.3	Cl	103 mEq/L	98 - 106	Ser Alb	3.0 g/dl	3.0 - 5.0
BUN	31 mg/dl	7 - 18	Cr	2.2 mg/dl	0.6 - 1.3	P	5.7 mg/dl	2.5 - 4.5
Uric Acid	8.8 mg/dl	3.5 - 7.2	Ca	7.7 mg/dl	8.6 - 10.0			

****

QUESTIONS:

1.   Explain why Mr. G had leg cramps and explain what he could do to prevent them.

2.   Explain the pathophysiology of renal disease and HTN.

3.   Look up the anatomy of the kidney and define the following terms:

Glomerulus:

Nephron:

Loop of Henle:

Medulla:

Cortex:

4.   Which lab values in Table I and which results of the urinalysis indicate that Mr. G has a renal problem?

5.   Which lab values indicate protein malnutrition? In your answer, give the possible reasons for protein malnutrition and explain the relationship between protein in the urine and renal disease.

6.   Indicate which lab values in Table I suggest that Mr. G is anemic. Describe the relationship between kidney failure and anemia.

****

The nephrologist told Mr. G that his kidneys were failing and that he would have to go on a very strict diet and medication schedule. Orders from the nephrologist were as follows:

- ▸ 1. Tenormin
- ▸ 2. Lasix
- ▸ 3. Phos-Lo
- • 4. Epogen
- ▸ 5. Multivitamin with minerals
- ▸ 6. 0.6 g of protein per kg of body weight
- ▸ 7. kcals per RD's recommendation
- ▸ 8. 2g Na
- ▸ 9. No K restriction
- ▸ 10. No fluid restriction

Diets can be ordered in terms of mg or mEq of Na and K. Therefore, it is necessary to be able to change from one to the other.

DOCTOR'S ORDERS

10/28/95 Tenormin 50mg PO qAM & if to 100 mg qAM p̄
Xd.

Lasix 40mg po qAM 20mg PO p̄ VIII hr.

Phos-Lo 500mg tid c̄ meals

Epogen 50u/kg SC 3X/wk

1 multivit/mise qd

0.6 g prot/kg

halr per RD

K̄ Na

∅ K restriction

∅ fluid restriction

JH

Above is a mock page of doctors orders. See if you can transcribe the orders.

****

7.   Give the function of the following medications:

Phos-Lo:

Lasix:

Epogen:

8.   What are the nutritional implications/complications of these drugs?

9.   Explain the rationale for Mr. G's diet. Include in your discussion the reasons for the following:
- 0.6 g of protein per kg of body weight
- 2 g sodium
- No K restriction
- No fluid restriction

****

The nephrologist had Mr. G see the renal dietitian before he went home. The RD asked him who did the cooking. He told her that his wife did all the cooking and grocery shopping. The RD said that she wanted to talk to Mr. G and his wife together and made an appointment for them. Mr. G came back with his wife and the dietitian instructed them on diet and cooking techniques. She also impressed upon them the importance of following the diet and medication orders. Mr. G finally realized the importance of the diet and was frightened. He remembered what happened to his uncle who had had total kidney failure. He went home declaring that he would follow his diet to the letter. The nephrologist told Mr. G that if he stayed on his diet and took his medications, his renal disease may not get any worse and he may not need dialysis, or he may be able to avoid it for years to come. The nephrologist also told him that if he did not follow his diet and did not take his medications, he would very likely end up with ESRD.

Mr. G followed his diet much more closely than he did last time and he also took all of his medications. The increased dosage of Lasix caused him to go to the bathroom more and he was losing some weight. This went on for several months without a problem and then he noticed that he was not going to the bathroom as much. The edema was coming back and so was the weight. His pain in the left lower flank had subsided but was now returning. Mr. G also noticed that he was feeling very tired and would get weak after doing easy tasks. He began to experience anorexia, headaches, and nausea. Mr. G went back to his nephrologist, who obtained another set of lab values. These lab values were as follows:

Table II

TEST	RESULT	NORM	TEST	RESULT	NORM	TEST	RESULT	NORM
Hgb	16.4 g/dl	14.7 - 17.4	MCV	83 μm³	82 - 98	Glucose	103 mg/dl	65 - 110
Hct	44%	42 - 52	Mg	1.9 mEq/L	1.3 - 2.1	Na	149 mEq/L	135 - 145
K	6.5 mEq/L	3.5 - 5.3	Cl	103 mEq/L	98 - 106	Ser Alb	3.4 g/dl	3.9 - 5.0
BUN	47 mg/dl	7 - 18	Cr	7.9 mg/dl	0.6 - 1.3	P	6.2 mg/dl	2.5 - 4.5
Uric Acid	9.3 mg/dl	2.6 - 6.0	Ca	7.8 mg/dl	8.6 - 10.0			

Urinalysis:

SG 1.001   Protein +1   Glucose neg.   Ketones neg.   pH 8.2

The nephrologist told Mr. G that the results of his tests were not good. His kidneys were going into ESRD and he would have to go on a hemodialysis machine for about 3 hrs 3x/wk. He told Mr. G that he would require minor surgery to have an A-V graft created in his arm to provide access to the dialysis machine. He explained what an A-V graft was. Mr. G's weight was now 215 lbs.
****

QUESTIONS CONTINUED:

10.   Define nephrologist.

11.   Describe the A-V graft for hemodialysis.

12.   Describe hemodialysis. In your description, tell what hemodialysis does and what the complications are.

13.   List the dietary principles for a patient on hemodialysis.

****

The MD told Mr. G that his new diet would be as follows:
- ‣ 1. 1.3 g of protein per kg of body weight
- ‣ 2. 2 g Na
- ‣ 3. 90 mEq K
- ‣ 4. Fluid intake = urine output +500 cc

Mr. G's new orders for medications were as follows:
- ▪ 1. Continue Phos-Lo
- ▪ 2. Continue antihypertensive
- ▪ 3. Continue Epogen
- ▪ 4. D/C Lasix
- ▪ 5. Continue multivitamin

****

14.   Convert 60 mEq K to mg (Appendix A, Table A - 5).

15.   Why did the protein in the diet order increase with ESRD?

****

Mr. G was admitted to the hospital again to have an A-V graft created in his left arm. While in the hospital, Mr. G received hemodialysis for the first time. His weight upon admission to the hospital was 215 lbs. This was listed as his wet weight. After dialysis, Mr. G was weighed again and weighed 210 lbs. This was listed as his dry weight.

The renal dietitian met with Mr. G and his wife again and explained the changes in his new diet. She emphasized the importance of increasing the protein and restricting sodium and fluid. She also told Mr. G that it was important for him to take in enough energy. To do this, it will now be necessary for him to eat more sweets and fats. She suggested that whenever he eats bread or toast, he should add margarine, and whenever he eats salad, he should add lots of low-sodium salad dressing. She also suggested that he eat more candy that did not have salted nuts in it. He should add sugar to coffee, tea, Kool Aid, etc.

****

QUESTIONS CONTINUED:

16.   Explain the difference between the wet weight and the dry weight.

17.   On the next page, create a SOAP note for Mr. G using all of the information since and including Table II. Use any pertinent information prior to Table II as his previous history (See Information Box 4-1, page 37).

PROGRESS NOTES

S:

O:

A:

P:

_____
_____
_____
_____
_____
_____
_____

18. Calculate Mr. G's BEE using the Harris-Benedict equation and the appropriate stress factor (Appendix D, Tables D - 1 and 5).

19. Mr. G's energy and protein requirements on hemodialysis are higher than his energy and protein requirements without hemodialysis. Explain why.

20. The RD encouraged Mr. G to increase his caloric intake by eating more fat and sugar. Comment on the advantages vs. disadvantages of doing this with a patient who has renal disease.

****

Mr. G's wife asked the MD how long Mr. G would have to stay on his diet. The MD replied that he would have to stay on his diet for the rest of his life. The only way that he could change his diet would be to have a successful kidney transplant. He discussed the possibility of a kidney transplant with Mr. G and his wife. Mr. G decided that he wanted to be put on the list to receive a donor kidney.

****

21. Using Mr. G's last plan of MNT (1.3 g of protein per kg, 2 g Na, 90 mEq K, and 500 cc fluid + urinary output), plan a day's menu using renal diet exchanges (your textbook, your local diet manual, or reference # 14 would be good places to obtain renal diet exchange information).

Related References

1. Adamson, J.W. & Eschbach, J.W. (1998). Erythropoietin for end-stage renal disease. *N. Engl. J. Med.* 27;339(9):625-7.

2. Bergstrom, J., Wang, T., & Lindholm, B. (1998). Factors contributing to catabolism in end-stage renal disease patients. *Miner. Electrolyte Metab.* 24(1):92-101.

3. Besarab, A., Bolton, W.K., Browne, J.K., Egrie, J.C., Nissenson, A.R., Okamoto, D.M., Schwab, S.J., & Goodkin, D.A. (1998). The effects of normal as compared with low hematocrit values in patients with cardiac disease who are receiving hemodialysis and epoetin. *N. Engl. J. Med.* 27;339(9):584-90.

4. Burdick, C.O. (1998). Prealbumin and prediction of survival in dialysis patients. *Am. J. Kidney Dis.* 31(1):195.

5. Chohan, N. Senior Editor. (1998). Nursing 98 Drug Handbook. Springhouse Corporation, Springhouse, Pennsylvania.

6. Cockram, D.B., Hensley, M.K., Rodreguez, M., Agarwal, G., Wennberg, A., Ruey, P., Ashbach, D., Hebert, L., & Kunan, R. (1998). Safety and tolerance of medical nutritional products as sole sources of nutrition in people on hemodialysis. *J. Ren. Nutr.* 8(1):25-33.

7. Fischbach, F.T. (1995). *A Manual of Laboratory & Diagnostic Tests.* 5th Ed. Philadelphia. J.B. Lippincott Company.

8. Kaysen, G.A. (1998). Albumin turnover in renal disease. *Miner. Electrolyte Metab.* 24(1): 55-63.

9. Kaufman, J.S., Reda, D.J., Fye, C.L., Goldfarb, D.S., Henderson, W.G., Kleinman, J.G., & Vaamonde, C.A. (1998). Subcutaneous compared with intravenous epoetin in patients receiving hemodialysis. Department of Veterans Affairs Cooperative Study Group On Erythropoietin in Hemodialysis Patients. *N. Engl. J. Med.* 339(9):578-83.

10. Leung, J. & Dwyer, J. (1998). Renal DETERMINE nutrition screening tools for the identification and treatment of malnutrition. *J. Ren. Nutr.* 8(2):95-106.

11. Matarese, L.E., & Gottschlich, M.M. (1998). *Contemporary Nutrition Support Practice: A Clinical Guide.* Philadelphia, Pennsylvania. W.B. Saunders.

12. Prendergast, A. & Fulton, F.L. (1997). Medical Terminology: A Text/Workbook. 4th Ed. Redwood City, California. Addison-Wesley Nursing.

13. Pronsky, Z.M. & Solomon, E. (1997). *Food-Medication Interactions.* 10th Ed. Phoenix, Arizona. Food-Medication Interactions, Publishers and Distributors.

14. Renal Dietetians Dietetic Practice Group of The American Dietetic Association. (1993). *National Renal Diet: Professional Guide.* Chicago, IL,

15. Tai, T.W.C., Chan, A.M.W., Cochran, C.C., Harbert, G., Lindley, J., & Cotton., J. (1998). Renal Dietitians' Perspective: identification, prevalence, and intervention for malnutrition in dialysis patients in Texas. *J. Ren. Nutr.* 8(4):188-98.

16. Tang, W.W., Stead, R.A., & Goodkin D.A. (1998). Effects of Epoetin alfa on hemostasis in chronic renal failure. *Am. J. Nephrol.* 18(4):263-73.

17. *Manual of Clinical Dietetics.* (1996). 5th Ed. Chicago, IL. The American Dietetic Association.

18. Whitney, E.N., Cataldo, C.B., & Rolfes, S.R. (1998). *Understanding Normal and Clinical Nutrition*, 5th Ed. West/Wadsworth.

CASE STUDY #39
KIDNEY TRANSPLANT

INTRODUCTION

This is the third case study in a series of cases concerning hypertension and renal disease. This study involves a kidney transplant and points out the significant changes in diet between a hemodialysis patient and a transplant patient. Mr. G has been following his diet and taking his medications as he should for eight months and has been doing well. A donor has been found for a transplant and Mr. G is now back in the hospital preparing to receive his new kidney.

SKILLS NEEDED

ABBREVIATIONS:

Knowledge of the following abbreviations is required in order to understand this case. You should learn these abbreviations before you begin to read the study.

BUN : blood urea nitrogen
Ca : calcium
Cl : chloride
Cr : creatinine
D/C : discontinue
g : gram
g/dl : grams per deciliter
Hct : hematocrit
HD : hemodialysis
Hgb : hemoglobin

K : potassium
MCV : mean corpuscular volume
mEq/L : milliequivalent per liter
Mg : magnesium
mg/dl : milligram per deciliter
Na : sodium
P : phosphorous
Ser Alb : serum albumin
YOM : year old male
3x/wk : three times per week
μm^3 : cubic microns

FORMULAS:

The formulas used in this case study include total caloric and protein requirements (Appendix D, Tables D - 1 and 5).

LABORATORY VALUES:

You will need to be able to interpret the nutritional significance of the following laboratory values for this case study: BUN, Cl, Cr, Hct, Hgb, K, MCV, Mg, Na, P, glucose, and Ser Alb (Appendix B).

MEDICATIONS:

Become familiar with the following medications before reading the case study. Note the diet-drug interactions, dosages and methods of administration, gastrointestinal tract reactions, etc.

1. Prednisone (prednisone); 2. Sandimmune (cyclosporin); 3. Phos-Lo (calcium acetate).

Mr. G has been doing well on his diet and medications for the past eight months. He has been receiving HD 3 x/wk for 4 hrs at a time. He has not had any complications but continues to lose weight. His dry weight is now 196 lbs. He is very much concerned about his weight loss. Mr. G is now in the hospital being prepared for a kidney transplant. His most recent lab values were as follows:

TEST	RESULT	NORM	TEST	RESULT	NORM	TEST	RESULT	NORM
Hgb	11.7 g/dl	14 - 17.4	MCV	75 μm³	82 - 98	Glucose	115 mg/dl	65 - 110
Hct	37%	42 - 52	Mg	2.0 mEq/L	1.3 - 2.1	Na	141 mEq/L	135 - 145
K	5.4 mEq/L	3.5 - 5.3	Cl	100 mEq/L	98 - 106	Ser Alb	2.7 g/dl	3.9 - 5.0
BUN	52 mg/dl	7 - 18	Cr	3.5 mg/dl	0.6 - 1.3	P	6.3 mg/dl	2.5 - 4.5
Uric Acid	8.0 mg/dl	3.5 - 7.2	Ca	7.7 mg/dl	8.6 - 10.0			

Mr. G is to receive a kidney transplant that was found through the hospital's computerized system. The new kidney is a cadaver transplant from a 30 YOM that was killed in an automobile accident. Mr. G went to surgery and the transplant was completed successfully. After surgery, Mr. G recovered well without incident. His new orders read as follows:

▸ 1. prednisone
▸ 2. cyclosporin
▸ 3. D/C Phos-Lo
▸ 4. high protein, low carbohydrate, 2 g Na diet
▸ 5. no K restriction
▸ 6. no fluid restriction

The renal RD again met with Mr. G and his wife and explained the changes in his diet. She told him that he now does not need to limit his protein intake. In fact, she encouraged him to take in a lot of protein. She also cautioned him to avoid sweets such as sugar, candy, soft drinks, and desserts. She recommended that he reduce fat intake in his diet and emphasized the importance of following the 2 g Na recommendation. At first, Mr. G was confused because this was so different from the HD diet of limited protein, high carbohydrate, high fat, and restricted K.
****

QUESTIONS:

1.   The values indicating protein malnutrition are a little lower than they were the last time. Discuss what factors could be responsible for this.

2.   Describe the action of prednisone. Discuss the nutritional implications.

3.   Describe the action of cyclosporin and discuss the nutritional implications.

4.   Research and describe the methods to give cyclosporin orally.

5.   What are the nutritional goals for a patient who has just had a kidney transplant?

6.   Calculate Mr. G's energy and protein needs and compare these values with his needs before dialysis and during dialysis. Show your work.

7.   If you were the RD, how would you explain to Mr. G the reasons for the change in his diet? (☞ *Hint: Consider Mr. G's medications.*)

****

Mr. G recovered as expected without complications. His renal function improved dramatically. He was not back to normal and never would be, but as long as he stayed on his diet and continued his medications, his dietary restrictions should not get any worse. He went home from the hospital, followed his diet, and took his medications. He returned to work and was very thankful to have renewed energy and the ability to adequately complete his job. After about five months, Mr. G's

weight increased to 216 lbs. He noticed that the weight gain was fat and not muscle. He weighed as much as he did before he was sick, but he did not have the muscle mass he had prior to his illness. He was concerned about this and returned to see his RD. Upon visual examination, the RD could see that Mr. G had an increase in abdominal fat and a very full round face. He appeared to be much healthier than he was when he was receiving HD, but it was obvious that he had gained a considerable amount of fat. The RD explained to Mr. G why this was taking place.
****

QUESTIONS CONTINUED:

8.   Explain what is happening to Mr. G.

9.   For a patient on prednisone, an increased amount of fat tissue and a round full face is not uncommon. What is this condition called?

10.   Define the following terms:

        Cadaver Transplant:

        Sibling Transplant:

▸ **ADDITIONAL OPTIONAL QUESTION** ◂

Tube Feeding Drill:

11. Suppose Mr. G became seriously ill prior to receiving his transplant, had to be hospitalized, and could only be fed by a feeding tube. Using the table below, compare several of the enteral nutritional supplements that would be appropriate under these conditions.

Product	Producer	Form	Cal/ ml	Non- pro Cal/g N	g/L			Na mg	K mg	mOsm /kg Water	Vol to meet RDAs in ml	g of fiber /L	Free water /L in ml
					Pro	CHO	Fat						

Related References

1. Chohan, N. Senior Editor. (1998). Nursing 98 Drug Handbook. Springhouse Corporation, Springhouse, Pennsylvania.

2. Fischbach, F.T. (1995). *A Manual of Laboratory & Diagnostic Tests*. 5th Ed. Philadelphia. J.B. Lippincott Company.

3. Leung, J. & Dwyer, J. (1998). Renal DETERMINE nutrition screening tools for the identification and treatment of malnutrition. *J. Ren. Nutr.* 8(2):95-106.

4. Matarese, L.E. & Gottschlich, M.M. (1998). *Contemporary Nutrition Support Practice: A Clinical Guide*. Philadelphia, Pennsylvania. W.B. Saunders.

5. Prendergast, A. & Fulton, F.L. (1997). Medical Terminology: A Text/Workbook. 4th Ed. Redwood City, California. Addison-Wesley Nursing.

6. Pronsky, Z.M. & Solomon, E. (1997). *Food-Medication Interactions*. 10th Ed. Phoenix, Arizona. Food-Medication Interactions, Publishers and Distributors.

7. Renal Dietetians Dietetic Practice Group of The American Dietetic Association. (1993). *National Renal Diet: Professional Guide*. Chicago, IL.

8. The American Dietetic Association. *Manual of Clinical Dietetics*. (1996). 5th Ed. Chicago, IL.

9. Whitney, E.N., Cataldo, C.B., & Rolfes, S.R. (1998). *Understanding Normal and Clinical Nutrition*, 5th Ed. West/Wadsworth.

Related References

1. Alterescu, K.B. (1985). The ostomy. What about special procedures? *Am. J. Nurs.* 85(12):1363-7.

2. Black, P. (1985). Stoma care. Drugs and diet. *Nurs. Mirror.* 161(11):26-8.

3. Duerksen, D.R., Nehra, V., Bistrian, B.R., & Blackburn, G.L. (1998). Appropriate nutritional support in acute and complicated Crohn's disease. *Nutrition.* 14(5):462-5.

4. Elishoov, H., Or, R., Strauss, N., & Engelhard, D. (1998). Nosocomial colonization, septicemia, and Hickman/Broviac catheter-related infections in bone marrow transplant recipients. A 5-year prospective study. *Medicine (Baltimore).* 77(2):83-101.

5. Ferguson, A., Glen, M., & Ghosh, S. (1998). Crohn's disease: nutrition and nutritional therapy. *Baillieres. Clin. Gastroenterol.* 12(1):93-114.

6. Fernandez-Banares, F., Cabre, E., Gonzalez-Huix, F., & Gassull, M.A. (1994). Enteral nutrition as primary therapy in Crohn's disease. *Gut.* 35(1 Suppl):S55-9.

7. Fischbach, F.T. (1995). *A Manual of Laboratory & Diagnostic Tests.* 5th Ed. Philadelphia. J.B. Lippincott Company.

8. Geerling, B.J., Badart-Smook, A., Stockbrugger, R.W., & Brummer, R.J. (1998). Comprehensive nutritional status in patients with long-standing Crohn's disease currently in remission. *Am. J. Clin. Nutr.* 67(5):919-26.

9. Griffiths, A.M., Ohlsson, A., Sherman, P.M., & Sutherland, L.R. (1995). Meta-analysis of enteral nutrition as a primary treatment of active Crohn's disease. *Gastroenterology.* 108(4):1056-67.

10. Hamilton, H. & Fermo, K. (1998). Assessment of patients requiring i.v. therapy via a central venous route. *Br. J. Nurs.* 7(8):451-4, 456-60.

11. Hunter, J.O. (1998). Nutritional factors in inflammatory bowel disease. *Eur. J. Gastroenterol. Hepatol.* 10(3):235-7.

12. Husain, A. & Korzenik, J.R. (1998). Nutritional issues and therapy in inflammatory bowel disease. *Semin. Gastrointest. Dis.* 9(1):21-30.

13. Kushner, R.F. (1992). Should enteral nutrition be considered as primary therapy in acute Crohn's disease? *Nutr. Rev.* 50(6):166-9.

14. Messori, A., Trallori, G., D'Albasio, G., Milla, M., Vannozzi, G., & Pacini, F. Defined formula diets versus steriods in the treatment of active Crohn's disease: a meta-analysis. *Scand. J. Gastroenterol.* 31(3):267-72.

15. Mingrone, G., Benedetti, G., Capristo, E., De Gaetano, A., Greco, A.V., Tataranni, P.A., & Gasbarrini, G. (1998). Twenty-four-hour energy balance in Crohn's disease patients: metabolic implications of steroid treatment. *Am. J. Clin. Nutr.* 67(1):118-23.

16. Moran, B.J., Sutton, G.L., & Karran, S.J. (1992). Clinical evaluation of percutaneous insertion and long-term usage of a new cuffed polyurethane catheter for central venous access. *Ann. R. Coll. Surg. Engl.* 74(6):426-9.

17. Nursing 98 Books. (1998). *Drug Handbook.* Springhouse, Pennsylvania. Springhouse Corporation.

18. O'Sullivan, M.A. & O'Morain, C.A. (1998). Nutritional therapy in Crohn's disease. *Inflamm. Bowel Dis.* 4(1):45-53.

19. Prendergast, A. & Fulton, F.L. (1997). *Medical Terminology: A Text/Workbook.* 4th Ed.

20. Pronsky, Z.M. & Solomon, E. (1998). *Food-Medication Interactions.* 10th Ed. Phoenix, Arizona. Food-Medication Interactions, Publishers and Distributors.

21. Sales, T.R., Torres, H.O., Couto, C.M., & Carvalho, E.B. (1998). Intestinal adaptation in short bowel syndrome without tube feeding or home parenteral nutrition: report of four consecutive cases. *Nutrition.* 14(6):508-12.

22. Whitney, E.N., Cataldo, C.B., & Rolfes, S.R. (1998). *Understanding Normal and Clinical Nutrition,* 5th Ed. West/Wadsworth.

CASE STUDY #41
CROHN'S DISEASE AND TPN

INTRODUCTION

This is a continuation of Case Study #40. This study is advanced and involves TPN and tube feedings as related to Crohn's disease.

SKILLS NEEDED

ABBREVIATIONS:

Knowledge of the following abbreviations is required in order to understand this case. You should learn these abbreviations before you begin to read the study.

AA : amino acids
amp : ampule
cc/h : cubic centimeters per hour
CHO : carbohydrate
D/C : discontinue
g/d : grams per day
$D_{50}W$: 50% dextrose in water
HN : high nitrogen
I.V. : intravenous
kg : kilogram

mOsm/kg : milliosmole per kilogram
MVI : multiple vitamin infusion (Appendix E)
NST : nutrition support team
ozs : ounces
po : by mouth
qd : every day
tab : tablet
TPN : total parenteral nutrition
Wk : week
2x : two times
3x : three times

LABORATORY VALUES:

Review the labs normally used to monitor TPN and I.V. lipids (see Case Study #35 for additional guidance).

FORMULAS:

The formulas used in this case study include simple percentages and the calculation of kilocalorie and protein needs to determine TPN formulation. Appendices A, D, and E will be helpful.

MEDICATIONS:

Become familiar with the following medications before reading the case study. Note the diet-drug interactions, dosages and methods of administration, gastrointestinal tract reactions, etc.
1. Panasol (prednisone); 2. Azulfidine (sulfasalazine).

An order has been written for the dietitian to calculate Mrs. M's energy and protein needs and to recommend a TPN mixture, flow rate, and diet. In the meantime, her nutrition orders were standard TPN at 50 cc/h. Standard TPN consist of: 500 cc of $D_{50}W$, 500 cc of 8.5% AA, standard electrolytes with 1 amp of MVI 9 + 3, and 9 cc of trace minerals qd (Appendix E). This is to be given via a Hickman/Broviac catheter. The energy and protein needs were calculated in Case Study #40.
****

QUESTIONS:

1.   Would you give Mrs. M I.V. lipids? Why or why not?

2.   Differentiate between lipid administration as a three-in-one admixture or as piggy back. Include the advantages, disadvantages, and precautions of each.

3.   What lab value(s) are used to monitor lipids?

4.   If you decided to give lipids, how much would you give and what flow rate would you use?

5.   For TPN, would you use the standard formula or create a new formula?   Give reasons for your answer, and regardless of which formula you use, list the constituents and amounts you would recommend on a daily basis (assume you are working at a hospital that allows individual formulation of TPN formulas if necessary).

6. For whichever formula you used in question 5, calculate the amount of TPN Mrs. M will need daily to meet her energy and protein requirements. Consider the lipids you are giving in your calculations.

7. Calculate the flow rate that is necessary to deliver the amount of TPN calculated in question 6 in a day.

8. Double check your answer by calculating the total amount of CHO, protein, and fat Mrs. M will receive with the formula and flow rates you determined. Compare these values with your original calculations in question 5. Show all work.

9. Would you add any minerals above the standard amount? If so, which ones and how much? Explain your answer.

10. List the complications of TPN.

11. List the labs that are used to monitor TPN and the frequency in which they should be checked.

LAB CHECK: DAILY   3 X PER WK   2 X PER WK   WEEKLY   OTHER

****

Mrs. M's ileostomy output slowed down after a few days of bowel rest. With TPN and close monitoring, her electrolytes were soon in normal range. Mrs. M started to feel stronger, so po feedings were restarted. The NST recommended the following diet prescription:

▸ 1. Restrict complex CHO to 150 g/d (excluding CHO in the supplement)
▸ 2. No sweets
▸ 3. No more than 4 ozs of fruit juice per day
▸ 4. No caffeine
▸ 5. Chew food extremely well
▸ 6. Restrict total fluids to 1500 cc
▸ 7. 50 g protein (excluding protein in the supplement)
▸ 8. Low fat (no more than 38 g excluding fat in the supplement)
▸ 9. No salt restriction
▸ 10. At least 2 potassium sources per day
▸ 11. Stress multivitamin/mineral tab qd
▸ 12. In addition to above, 2 servings of a supplemental tube feeding appropriate for SBS. The dietitian can choose the appropriate supplement.

****

QUESTIONS CONTINUED:

12. Research the available supplements and suggest an appropriate supplement for SBS. In your choice, recommend an amount to be given twice per day. Provide the following information on the supplement (see Appendix F):

Nutrient	g/l	Source of Nutrient
CHO		
Protein		
Fat		
kcals/cc =		
mOsm/kg of water =		
amount of free water =		

13.   Research the medications prednisone and sulfasalazine. Tell what they are used for and list any nutritional implications.

14.   Briefly give the rationale for each of the recommendations made by the NST.
1.   Restrict complex CHO to 150 g/d (excluding CHO in the supplement)
2.   No sweets
3.   No more than 4 ozs of fruit juice per day
4.   No caffeine
5.   Chew food extremely well
6.   Restrict total fluids to 1500 cc (excluding liquid in the supplement)
7.   50 g protein (excluding protein in the supplement)
8.   Low fat (no more than 38 g, excluding fat in the supplement)
9.   No salt restriction
10.   At least 2 potassium sources per day
11.   Stress multivitamin/mineral tab qd

This diet was attempted by starting slowly and gradually building up to the amount prescribed. The TPN was decreased gradually until it was no longer needed. The pt tolerated the diet and was D/C without TPN. Her medications included prednisone po and sulfasalazine po. Her diet was changed slightly to include 200 g of CHO and 1 serving of supplement per day.

15.   For the diet that Mrs. M is being discharged with, calculate:
   1.   total kcals and kcals/kg
   2.   total grams of CHO
   3.   total protein
   4.   total fat
   5.   grams of protein/kg of body weight
   6.   stress factor used
   7.   ccs per kcal (fluid allowed plus supplement)

► **ADDITIONAL OPTIONAL QUESTION** ◄

Tube Feeding Drill:

16.   Using the table below, compare several of the enteral nutritional supplements that are comparable to the one you chose above.

Product	Producer	Form	Cal/ ml	Non- pro Cal/g N	g/L			Na mg	K mg	mOsm /kg Water	Vol to meet RDAs in ml	g of fiber /L	Free water /L in ml
					Pro	CHO	Fat						

Related References

1. Alterescu, K.B. (1985). The ostomy. What about special procedures? *Am. J. Nurs.* 85(12):1363-7.

2. Black, P. (1985). Stoma care. Drugs and diet. *Nurs. Mirror.* 161(11):26-8.

3. Duerksen, D.R., Nehra, V., Bistrian, B.R., & Blackburn, G.L. (1998). Appropriate nutritional support in acute and complicated Crohn's disease. *Nutrition.* 14(5):462-5.

4. Elishoov, H., Or, R., Strauss, N., & Engelhard, D. (1998). Nosocomial colonization, septicemia, and Hickman/Broviac catheter-related infections in bone marrow transplant recipients. A 5-year prospective study. *Medicine (Baltimore).* 77(2):83-101.

5. Ferguson, A., Glen, M., & Ghosh, S. (1998). Crohn's disease: nutrition and nutritional therapy. *Baillieres. Clin. Gastroenterol.* 12(1):93-114.

6. Fernandez-Banares, F., Cabre, E., Gonzalez-Huix, F., & Gassull, M.A. (1994). Enteral nutrition as primary therapy in Crohn's disease. *Gut.* 35(1 Suppl):S55-9.

7. Fischbach, F.T. (1995). *A Manual of Laboratory & Diagnostic Tests.* 5th Ed. Philadelphia. J.B. Lippincott Company.

8. Geerling, B.J., Badart-Smook, A., Stockbrugger, R.W., & Brummer, R.J. (1998). Comprehensive nutritional status in patients with long-standing Crohn's disease currently in remission. *Am. J. Clin. Nutr.* 67(5):919-26.

9. Griffiths, A.M., Ohlsson, A., Sherman, P.M., & Sutherland, L.R. (1995). Meta-analysis of enteral nutrition as a primary treatment of active Crohn's disease. *Gastroenterology.* 108(4):1056-67.

10. Hamilton, H. & Fermo, K. (1998). Assessment of patients requiring i.v. therapy via a central venous route. *Br. J. Nurs.* 7(8):451-4, 456-60.

11. Hunter, J.O. (1998). Nutritional factors in inflammatory bowel disease. *Eur. J. Gastroenterol. Hepatol.* 10(3):235-7.

12. Husain, A. & Korzenik, J.R. (1998). Nutritional issues and therapy in inflammatory bowel disease. *Semin. Gastrointest. Dis.* 9(1):21-30.

13. Kushner, R.F. (1992). Should enteral nutrition be considered as primary therapy in acute Crohn's disease? *Nutr. Rev.* 50(6):166-9.

14. Messori, A., Trallori, G., D'Albasio, G., Milla, M., Vannozzi, G., & Pacini, F. Defined formula diets versus steroids in the treatment of active Crohn's disease: a meta-analysis. *Scand. J. Gastroenterol.* 31(3):267-72.

15. Mingrone, G., Benedetti, G., Capristo, E., De Gaetano, A., Greco, A.V., Tataranni, P.A., & Gasbarrini, G. (1998). Twenty-four-hour energy balance in Crohn's disease patients: metabolic implications of steroid treatment. *Am. J. Clin. Nutr.* 67(1):118-23.

16. Moran, B.J., Sutton, G.L., & Karran, S.J. (1992). Clinical evaluation of percutaneous insertion and long-term usage of a new cuffed polyurethane catheter for central venous access. *Ann. R. Coll. Surg. Engl.* 74(6):426-9.

17. Nursing 98 Books. (1998). *Drug Handbook.* Springhouse, Pennsylvania. Springhouse Corporation.

18. O'Sullivan, M.A. & O'Morain. (1998). Nutritional therapy in Crohn's disease. *Inflamm. Bowel Dis.* 4(1):45-53.

19. Prendergast, A. & Fulton, F.L. (1997). *Medical Terminology: A Text/Workbook.* 4th Ed.

20. Pronsky, Z.M. & Solomon, E. (1998). *Food-Medication Interactions.* 10th Ed. Phoenix, Arizona. Food-Medication Interactions, Publishers and Distributors.

21. Sales, T.R., Torres, H.O., Couto, C.M., & Carvalho, E.B. (1998). Intestinal adaptation in short bowel syndrome without tube feeding or home parenteral nutrition: report of four consecutive cases. *Nutrition.* 14(6):508-12.

22. Whitney, E.N., Cataldo, C.B., and Rolfes, S.R. (1998). *Understanding Normal and Clinical Nutrition,* 5th Ed. West/Wadsworth.

CASE STUDY #42
BURNS

INTRODUCTION

This study examines the nutritional treatment of a patient who receives severe burns to a significant percent of his body surface area. The student should review all aspects of stress and trauma, i.e., assessment, treatment, monitoring, and teaching the patient after he has experienced the initial trauma. Many terms are presented and TPN and tube feedings are involved with the assessment and monitoring of laboratory values.

SKILLS NEEDED

ABBREVIATIONS:

Knowledge of the following abbreviations is required in order to understand this case. You should learn these abbreviations before you begin to read the study.

BCAA : branched chain amino acids
BEE : basal energy expenditure
BSA : body surface area
BS : bowel sounds
C : centigrade
Ca : calcium
cc/h : cubic centimeters per hour
CHO : carbohydrate
D/C : discontinue
D_5W : 5% dextrose in water
F : Fahrenheit
g : gram
GI : gastrointestinal
g/dl : grams per deciliter
ICU : intensive care unit
I.V. : intravenous
K : potassium
KUB : kidney, ureter, and bladder (x-ray)
L : liter
mcg/dl : microgram per deciliter

mEq : milliequivalent
mg/dl : milligram per deciliter
Mg : magnesium
MS : morphine sulfate
N : nitrogen
N/G : nasogastric
OR : operating room
P : phosphorus
po : by mouth
qd : every day
RBW : reference body weight
Ser Alb : serum albumin
TF : tube feeding
TPN : total parenteral nutrition
UUN : urinary urea nitrogen
WBC : white blood count
YOBM : year old black male
1st° : first degree
2nd° : second degree
3 rd° : third degree

LABORATORY VALUES:

You will need to be able to interpret the nutritional significance of the following laboratory values for this case study: Ca, glucose, Mg, P, Ser Alb, WBC, UUN, K, and transferrin (Appendix B).

FORMULAS:

The formulas used in this case study include conversion of centigrade to Fahrenheit, reference body weight, percent reference body weight, total energy expenditure using the Harris-Benedict equation, UUN determinations, an appropriate stress factor, and rule of nines (Appendices A and

D, Tables A - 4, 7 through 10, 16, 17, and D - 1, 4, and 5).

MEDICATIONS:

Become familiar with the following medications before reading the case study. Note the diet-drug interactions, dosages and methods of administration, gastrointestinal tract reactions, etc.

1. Silvadene (silver sulfadiazine); 2. morphine sulfate; 3. Nafcil (naficillin sodium); 4. Lactated Ringer's solution.

AW is a 35 YOBM who was involved in an accident in his home that resulted in the explosion of a kerosene heater. AW attempted to move a kerosene heater a short distance without turning it off. The heater tipped over, spilling and igniting the fuel. AW received 1st°, 2nd°, and 3rd° burns over approximately 35 percent of his BSA. He was rushed to the local burn center, placed in ICU, and immediately intubated. A Foley catheter was inserted and a N/G tube was placed to low suction. A central line was inserted and fluid replacement therapy started using lactated Ringer's solution. AW was treated with the following medications:

▸   1. silver sulfadiazine (Silvadene)
▸   2. MS
▸   3. nafcillin sodium (Nafcil)

Before his accident, AW was 5'11" and weighed 185 lbs. He was a professional mover and was in reasonably good physical shape. He also jogged on occasion. In the burn unit, his room was kept at a temperature of 32° C and he was placed in protective isolation.
****

QUESTIONS:

1.   Convert 32° C to F (Appendix A, Table A - 4).

2.   Why was AW's room kept at a temperature of 32° C?

3.   Determine AW's RBW and percent RBW.

4.   Determine AW's energy and protein needs using the appropriate formulas for burn patients (Appendix D, Tables D - 1, 4, and 5). Show all work.

****

AW's burns covered the front of both legs, his abdomen, and some of his chest. His arms were burned in the front and a few small spots on his face were burned. The fire actually took place after the heater tilted and he had jumped back a few feet. It was estimated by the rule of nines that about 35 percent of BSA was burned. In some places his clothes had burned and had adhered to his body. This left deep wounds that oozed considerably. His dressing changes were very painful. A tracheostomy had to be done and AW was connected to a respirator. His urinary output decreased considerably and was only 60 cc per hour but this output was adequate. Because of his extensive burns, he could not be weighed nor could anthropometric measurements be obtained. Most of AW's fluid replacement came from I.V. lactated Ringer's solution, but he was also receiving D_5W at 65 cc/h and several units of salt-poor albumin qd. His abdomen was distended and hard to touch. There were no BS. Sterile technique was used for all contact with AW. TPN was started on the fourth day post-burn. Before starting TPN, some lab values were obtained as follows:

TEST	RESULT	NORM	TEST	RESULT	NORM	TEST	RESULT	NORM
Ser Alb	2.4 g/dl	3.9 - 5.0	P	2.0 mg/dl	2.5 - 4.5	Ca	7.5 mg/dl	8.6 - 10.0
Glucose	182 mg/dl	65 - 110	Trans-ferrin	170 mcg/dl	200 - 400	Mg	1.1 mEq/L	1.3 - 2.1
K	3.1 mEq/L	3.5 - 5.3	WBC	$5.0x$ $10^3/mm^3$	$5 - 10 x$ 10^3	24 hr UUN	18 g N	-

These labs were collected early in the morning of the fourth day post-burn before the TPN solution was started. Urinary output had increased to 75 cc per hour when the UUN was obtained. There were no signs of infection. AW still had a N/G tube to low suction.
****

QUESTIONS CONTINUED:

5.   Give the mechanism of action for the following drugs:

Silver sulfadiazine (Silvadene):

Morphine sulfate:

Nafcillin sodium (Nafcil):

6. List any adverse reactions or diet-drug interactions of the following drugs:

    Silver sulfadiazine (Silvadene):

    Morphine sulfate:

    Nafcillin sodium (Nafcil):

7. Describe the rule of nines (Appendix A, Table A - 16).

8. AW was placed in protective isolation. What does that mean? BS were listened to by sterile technique. What does that mean?

9. Since there are no anthropometric measurements for AW, how would you monitor his nutritional progress?

10. Explain the theory behind giving salt-poor albumin.

11. Compare the nutrient make-up of lactated Ringer's solution with that of extracellular fluid.

12.   Describe the test for UUN in respect to:
        a) How is the test conducted?

        b) What are the standards used to interpret the results?

        c) What interfering factors or conditions could affect this test?

13.   AW had 18 g of N in his urine by day four post-burn. How much protein does that represent? Show your work. (Appendix A, Table A - 17).

14.   AW's urinary output is 75 cc per hr. What are the other sources of fluid loss?

15.   AW's Ser Alb and transferrin were low and his UUN was high. How reliable are these two values given the above circumstances? Explain your answer in each case. (☞Hint: consider the loss of body fluids through oozing).

16.   Estimate the total protein loss that AW was having each day, including the UUN loss, and the loss through the BSA (Appendix A, Table A - 17). Show work.

17.   Is there a method to measure the amount of protein loss through the oozing of burns? If so, describe the method.

18.   What is the best estimation of AW's energy and protein needs under the above mentioned conditions?

19.   Has AW's energy and protein need changed from day one to day four post-burn? If so, indicate what this change would be and explain why.

20.   Branched chain amino acids are frequently used for burn patients. What is the theoretical advantage of this? Would you use BCAA for AW? Why or why not?

21. List the branched chain amino acids.

22. Complete the following calculations for AW's TPN:
a) The kind and amount of protein that should be used per L.

b) The amount of CHO that AW is receiving from D_5W in grams and calories.

c) The amount of CHO that should be used per L of TPN in addition to the amount received from the D_5W.

d) What vitamins, minerals, and electrolytes would you add daily? Are there any vitamins and/or minerals that should be added in pharmacological doses?

e) If you would use lipids, indicate the percent and amount per day. If you do not use lipids, explain why not.

f) How would you monitor AW's TPN; i.e., what lab values would you obtain and how often should they be repeated?

g) By approximately what day would you hope to be meeting AW's needs for energy and protein via the TPN solution?

23.   AW was not fed until the fourth day. This is presently an outdated method of feeding burn patients. According to the current literature, what is a preferred method of feeding the burn patient? In answering this question, include the goals of nutritional therapy you would have set for AW on his admission day in the emergency room.

****

By the tenth day post-burn, AW was doing well. He had been to the OR twice for skin grafts and debridement and had been to the OR for a third time for debridement. His urinary output was almost back to normal and his oozing had decreased a significant amount. His UUN on the seventh day was 20 g of N and his Ser Alb was holding steady at 2.0 g/dl. His transferrin was 150 mcg/dl. There were no signs of infection and the doctors were pleased with AW's progress since his nitrogen loss and BEE should now be peaking. He was tolerating his TPN well. P, Ca, and Mg were now in their normal ranges. AW was starting to have some BS and his abdomen was not as distended. He was still having considerable pain with dressing changes and was requiring MS. His dependency on a respirator continued because he had just come back from surgery. It was anticipated that he would be weaned in the next 2 to 3 days. The physicians wanted to start tube feedings, but they were concerned because they did not want AW to have diarrhea. However, to prevent the small bowel from becoming atrophic, they placed a feeding tube into the duodenum. They emphasized the importance of the correct placement of the feeding tube into the small bowel and confirmed its placement with a KUB. Once the feeding tube was properly positioned, they started half-strength fruit juice via the feeding tube at 15 cc/hr.

****

QUESTIONS CONTINUED:

24.   Why is it important for AW not to have diarrhea?

25.   What is the basis for the MD's concern about the small bowel becoming atrophic? Would fruit juice help prevent this? If so, how?

26.   After attempting to place a feeding tube into the small bowel, a KUB is usually obtained to confirm the correct placement. It may take several attempts with several KUBs before prpoer placement is obtained. Why is this so important?

****

By the twentieth day post-burn, AW's condition had improved. He returned to surgery for more skin grafts that were succesful. His UUN on the fourteenth day decreased to 8, his Ser Alb increased to 2.4 g/dl, and his transferrin was 200 mcg/L. A diluted tube feeding was being administered at a slow rate.

****

QUESTIONS CONTINUED:

27.   Discuss the type of tube feeding you would recommend for AW. Include in your discussion your recommendation for the initial flow rate and strength and the goal you would set for the final rate and strength.

28.   At some point, AW would have both a tube feeding and a TPN solution infusing at the same time. The goal should be to discontinue the TPN solution and rely on the tube feeding. Describe how you would complete the transition from the TPN solution to the tube feeding. Would you stop the TPN solution and start the tube feeding? Would you gradually slow down the TPN solution while you increased the tube feeding? Explain your answer and give the changes in the flow rates of each feeding method you would recommend.

29.   What criteria would you use in making your decision to change from TPN to TF?

30.   Assume BCAA were used early in AW's treatment. What criteria would you use when deciding to change from BCAA to another form of protein? Which form of protein would you change to?

31.   Would you recommend that AW continue receiving a vitamin or mineral supplement? If so, explain which ones and why.

****

Soon after the tube feeding was started, AW's endotracheal tube was removed. Diarrhea did not occur. AW continued to heal and gain strength so po feedings were begun and the TF D/C'd. AW is now approaching a positive nitrogen balance. Even though AW is doing well, he is very depressed about his appearance. He is most concerned about his face and keeps asking questions like "What does my face look like to you?" Although he was provided with a mirror to see his face (which was not badly burned), he continues to ask each day, "What does my face look like today?" His concern has affected his appetite. All those who attend to him encourage him to eat. His po intake is not as good as the intake he was receiving with the tube feeding. He has not healed completely and still requires several skin grafts.

****

QUESTIONS CONTINUED:

32.   What are the nutritional goals for AW's po intake? Include in your discussion any vitamin or mineral supplements that should be considered.

33. Identify the following terms:

    Intubated:

    Trach:

    Tracheostomy:

    Respirator:

    Skin Grafts:

    Debridement:

    Ventilator Weaned:

    Atrophic:

    Foley:

    N/G to low suction:

► **ADDITIONAL OPTIONAL QUESTIONS** ◄

Tube Feeding Drill:

34. If your hospital had access to a metabolic cart, discuss how you could incorporate its use into AW's medical nutrition therapy.

35. In answering question 27, you recommended a type of tube feeding for AW. Using the table below, compare several of the enteral feedings available that would be appropriate for AW. Include the feeding you recommended in question 27.

Product	Producer	Form	Cal/ml	Non-pro Cal/g N	g/L			Na mg	K mg	mOs m/kg Water	Vol to meet RDAs in ml	g of fiber /L	Free water /L in ml
					Pro	CHO	Fat						

Related References

1.   Centinkale, O., & Yazici, Z. (1997). Early postburn fatty acid profile in burn patients. *Burns.* 23(5):392-9.

2.   Chuntrasakul, C., Siltharm, S., Sarasombath, S., Sittapirochana, C., Leowattana, W., Chockvivatanavanit, S. & Bunnak, A. (1998). Metabolic and immune effects of dietary arginine, glutamine and omega-3 fatty acids supplementation in immunocompromised patients. *J. Med. Assoc. Thai.* 81(5):334-43.

3.   Demling, R.H. & DeSanta, L. (1998). Increased protein intake during the recovery phase after severe burns increases body weight gain and muscle function. *J. Burn Care Rehabil.* 19(2):160-8.

4.   De-Souza, D.A. & Greene, L.J. (1998). Pharmacological nutrition after burn injury. *J. Nutr.* 128(5):797-803.

5.   Dhanraj, P., Chacko, A., Mammen, M., & Bharathi, R. (1997). Hospital-made diet versus commercial supplement in postburn nutritional support. *Burns.* 23(6):512-4.

6.   Fischbach, F.T. (1995). *A Manual of Laboratory & Diagnostic Tests.* 5th Ed. Philadelphia. J.B. Lippincott Company.

7.   Fratianne, R.B. & Brandt, C.P. (1997). Improved survival of adults with extensive burns. *J. Burn Care Rehabil.* 18(4):347-51.

8.   Hansbrough, J.F. (1998). Enteral nutritional support in burn patients. *Gastrointest. Endosc. Clin. N. Am.* 8(3):645-67.

9.   LeBoucher, J., Coudray-Lucas, C., Lasnier, E., Jardel, A., Ekindijian, O.G., & Cynober, L.A. (1997). Enteral administration of ornithine alpha-ketoglutarate or arginine alpha-ketoglutarate: a comparative study of their effects on glutamine pools in burn-injured rats. *Crit. Care Med.* 25(2):293-8.

10. LeBricon, T., Coooudray-Lucas, C., Lioret, N., Lim, S.K., Plassart, F, Schlegel, L., DeBandt, J.P., Saizy, R., Giboudeau, J., & Cynober, L. (1997). Ornithine alpha-ketoglutarate metabolism after enteral administration in burn patients: bolus compared with continuous infusion. *Am. J. Clin. Nutr.* 65(2):512-8.

11. Long, C.L., Schaffel, N., Geiger, J.W., Schiller, W.R., & Blakemore, W.S. (1979). Metabolic response to injury and illness: estimation of energy and protein needs from indirect calorimetry and nitrogen balance. *J. Parenter Enteral Nutr.* 3(6):452-456.

12. Mayes, T. (1997). Enteral nutrition for the burn patient. *Nutr. Clin. Pract.* 12(1 Suppl): S43-5.

13. Matarese, L.E. & Gottschlich, M.M., Editors. (1998). *Contempory Nutrition Support Practice: A Clinical Guide.* Philadelphia, Pennsylvania. W.B. Saunders.

14. Mayes, T., Gottschlich, M.M., & Warden, G.D. (1997). Clinical nutrition protocols for continuous quality improvements in the outcomes of patients with burns. *J. Burn Care Rehabil.* 18(4)L364-8.

15. Miller, S.F., Finley, R.K., Waltman, M., & Lincks, J. (1991). Burn size estimate reliability: a study. *J. Burn Care Rehabil.* 12(60):546-559.

16. Nursing 98 Books. (1998). *Drug Handbook.* Springhouse, Pennsylvania. Springhouse Corporation.

17. Prendergast, A. & Fulton, F.L. (1997). *Medical Terminology: A Text/Workbook.* 4th Ed.

18. Pronsky, Z.M. & Solomon, E. (1998). *Food-Medication Interactions.* 10th Ed. Phoenix, Arizona. Food-Medication Interactions, Publishers and Distributors.

19. Raff, T., Hartmann, B., & Germann, G. (1997). Early intragastric feeding of seriously burned and long-term ventilated patients: a review of 55 patients. *Burns.* 23(1):19-25.

20. Skipper, A., Editor. (1998). *Dietitian's Handbook of Enteral and Parenteral Nutrition.* 2nd Ed. Gaithersburg, Maryland. ASPEN Publishers.

21. Whitney, E.N., Cataldo, C.B., & Rolfes, S.R. (1998). *Understanding Normal and Clinical Nutrition,* 5th Ed. West/Wadsworth.

CASE STUDY #43
PERITONEAL DIALYSIS

INTRODUCTION
This study examines the nutritional treatment of a patient with end stage renal disease who is receiving peritoneal dialysis. The student should review the medical nutrition therapy for the various types of dialysis treatments prior to studying this case.

SKILLS NEEDED

ABBREVIATIONS:
Knowledge of the following abbreviations is required in order to understand this case. You should learn these abbreviations before you begin to read the study.

BUN : blood urea nitrogen
Ca : calcium
CAPD : continuous ambulatory peritoneal dialysis
cm : centimeter
Cr : creatinine
ESRD : end stage renal disease
g : gram
g/dl : grams per deciliter
HD : hemodialysis
ht : height
K : potassium
kcals : kilocalories
kg : kilogram

L : liter
mEq : milliequivalent
Mg : magnesium
mg : milligram
mg/dl : milligram per deciliter
ml : milliliter
Na : sodium
P : phosphorus
phos : phosporous
PD : peritoneal dialysis
RBW : reference body weight
Ser Alb : serum albumin
YOWM : year old white male

LABORATORY VALUES:
You will need to be able to interpret the nutritional significance of the following laboratory values for this case study: Ca, glucose, Mg, P, Ser Alb, K, Na, BUN, and Cr (Appendix B).

FORMULAS:
The formulas used in this case study include metric conversions, reference body weight (Appendix A, Tables A -1, 2, and 7 through 9 and Information Box 43 - 1) and energy expenditure for renal patients (Appendix D, Table D - 5).

MEDICATIONS:
Become familiar with the following medications before reading the case study. Note the diet-drug interactions, dosages and methods of administration, gastrointestinal tract reactions, etc.
1. Phos-Lo (calcium acetate); 2. Epogen (epoetin alfa); 3. Calcijex (calcitriol); 4. Tenormin (atenolol).

BT is a 45 YOWM who is a successful executive and has a history of malignant hypertension. Because of his lifestyle, it was not convenient for him to comply with his diet and medication regimens. As a result, BT was diagnosed with ESRD and required dialysis. Hemodialysis was not an option for him because it required that he sit still for three hours three times a week. He is a very busy man and did not have time for hemodialysis. When continuous ambulatory peritoneal dialysis (CAPD) was described to him, it seemed like his best option. A catheter was surgically placed into his peritoneal space and he was instructed on how to complete the procedure of CAPD. BT has been on CAPD for three years.

Periodically, BT meets with the dietitian for an evaluation. In his last evaluation the following information was obtained from his medical record:

BT's ht is 177.8 cm and he weighs 75 kg. His usual body weight prior to renal failure was 70 kg. His daily CAPD dialysate prescription is as follows: 2 exchanges of 4.25% dextrose alternating with 2 exchanges of 1.5% dextrose, each exchange to dwell in his peritoneum for 4 hrs during waking hours; 1 exchange of 2.5% dextrose to dwell in his peritoneum for 6 hrs during sleeping hours. An exchange consist of 2 L.

His lab values were as follows:

TEST	RESULT	NORM	TEST	RESULT	NORM	TEST	RESULT	NORM
Ser Alb	4.2 g/dl	3.9 - 5.0	P	6.2 mg/dl	2.5 - 4.5	Ca	8.6 mg/dl	8.6 - 10.0
Glucose	100 mg/dl	65 - 110	Na	133 mEq/L	134 - 145	Mg	1.5 mEq/L	1.3 - 2.1
K	4.8 mEq/L	3.5 - 5.3	BUN	58 mg/dl	7 - 18	Cr	7 mg/dl	0.6 - 1.3

His nutritional prescription included the following: 35 kcals/kg of RBW, 1.2 g of protein per kg of RBW, 3 g of Na, K unrestricted, 2000 ml of fluid + urinary output, Ca supplemented by medication, and 15 mg of Phos per kg of RBW. The dietitian discussed BT's diet with him and obtained a 24 hour recall. She was convinced that BT was complying with his diet and medication plans reasonably well. Upon her evaluation, she recommended reducing his kcals to 25/kg of RBW, increasing his protein to 1.4 g/kg of RBW, increasing Na to 4g, increasing his calcium medication (calcium acetate), and letting K, Phos, and fluid remain the same.
****

QUESTIONS:

1.   Define the following terms:

        Malignant hypertension:

        Dialysate:

        Continuous Peritoneal Ambulatory Dialysis:

2.   Convert BT's height and weight from metric to English and determine his RBW (Appendix A, Tables A - 1, 2, and 7 through 9).

3.   Calculate the kcals and grams of protein, Na, and phos BT should be consuming in the original diet plan and the new plan recommended by the dietitian.

	Original Plan	New Plan
Kcals		
Protein		
Na		
Phos		

****

BT's medications included:
- Phos-Lo (calcium acetate)
- Epogen (epoetin alfa)
- Calcijex (calcitriol)
- Tenormin (atenolol)

****

Information Box 43 - 1

Peritoneal dialysis effectively removes waste products from the blood by making use of the semipermeable membrane of the peritoneum. After the surgical implantation of a catheter into the peritoneal cavity, a concentrated dialysate containing dextrose is infused into the peritoneum. Waste products diffuse from the blood across the peritoneum into the dialysate. This method is effective but not perfect. Substances other than waste products also diffuse into the dialysate, such as amino acids. Also, since the concentration of glucose in the dialysate is higher than the concentration of glucose in the blood, glucose diffuses across the peritoneum into the blood. This can be a significant amount of glucose and has to be accounted for. The volume of the dialysate and concentration of glucose in the dialysate varies from treatment to treatment. This makes it difficult to place a constant value on the amount of glucose that may infuse into the blood per treatment. The following formula is used to estimate the calories obtained from peritoneal dialysis:

$Y = [(11.3X) - 10.9] \times L$ of dialysate x 3.7 kcal/g of glucose where:

Y = grams of glucose absorbed per liter of dialysate

X = the concentration of glucose in g/L

L = a liter

1g of glucose = 3.7 kcals

The possible concentrations of glucose in dialysate include:

1.5% dextrose dialysate = 1.3 g of glucose/dL absorbed
2.5% dextrose dialysate = 2.2 g glucose/dL absorbed
4.25% dextrose dialysate = 3.8 g glucose/dL absorbed

Example: If the dialysate prescription consisted of 3 exchanges of 1.5%, 2 exchanges of 2.5%, and 3 exchanges of 4.25% glucose, what would be the amount of kcals diffused from the dialysate? Each exchange is 2 L.

1. Determine the grams of glucose absorbed from the various exchanges by multiplying the number of exchanges by the liters per exchange times the glucose concentration.

3 exchanges x 2L x 1.3 g/dl = 7.8 g
2 exchanges x 2L x 2.2 g/dl = 8.8 g
3 exchanges x 2L x 3.8 g/dl = 22.8 g

2. Add the total grams of glucose diffused.

Total = 7.8 + 8.8 + 22.8 = 34.4 g

3. Determine the grams of glucose per liter (X) by dividing the total grams by the total liters.

34.4 ÷ 16 = 2.15 g glucose/L

```
┌─────────────────────────────────────────────────────────────────────┐
│                        Information Box 43 - 1                          │
├─────────────────────────────────────────────────────────────────────┤
│ 4.  Plug X into the formula                                           │
│                                                                       │
│ Y = [11.3 x 2.15 - 10.9] x 16 L x 3.7 kcal/g glucose =                │
│    = 13.4 x 59.2                                                      │
│    = 793 kcals absorbed                                               │
│                                                                       │
│ For further explanation, see reference                                │
└─────────────────────────────────────────────────────────────────────┘
```

4.   Calculate the kcals BT is receiving from the dialysate described in his prescription.

5.   Why is the caloric value for glucose in the above formula 3.7 kcals per gram instead of the usual 4 kcals per gram?

6.   Usually the amounts of dietary protein, Na, K, and fluids allowed on CAPD are more than what is allowed for HD and PD. The kcal allotment for CAPD is usually lower than HD and PD. Explain why this is so.

7. If BT was following his diet and medication plans as prescribed, the dietitian was making her determinations based on his anthropometric measurements and lab values. Using these two parameters, give possible explanations for the changes the dietitian recommended in BT's nutrition care plan.

    Kcals:

    Protein:

    Na:

    Ca:

8. Why would she recommend K, phos, and fluid remain the same?

9. Compare the differences between hemodialysis, peritoneal dialysis, and continuous ambulatory peritoneal dialysis as related to kcals, protein, Na, K, Phos, and fluid intake.

	HD	PD	CAPD
Kcals			
Protein			
Na			
K			
Phos			
Fluid			

10.   List the advantages/disadvantages of hemodialysis, peritoneal dialysis, and continuous ambulatory peritoneal dialysis.

	Advantages	Disadvantages
HD		
PD		
CAPD		

11.   It is recommended that at least 50% of the protein fed to PD patients be of HBV. What does that mean and why is it important?

12.   What is the pharmacological classification of BT's drugs?

Phos-Lo:

Epogen:

Calcijex:

Tenormin:

13.   Describe the relationships of Phos-Lo, Epogen, and Calcijex to ESRD.

Phos-Lo:

Epogen:

Calcijex:

14. Another medical problem that patients with ESRD have to contend with is renalosteodystrophy. Relate why this is a concern.

15. Calcium acetate is the most effective phosphate binder for ESRD patients. Aluminum hydroxide gels used to be the treatment of choice years ago. Explain why these gels are no longer recommended for ESRD.

Related References

1. Adamson, J.W. & Eschbach, J.W. (1998). Erythropoietin for end-stage renal disease. *N. Engl. J. Med.* 27;339(9):625-7.

2. Bannister, D.K., Acchiardo, S.R., Moore, L.W., & Kraus, A.P. Jr. (1987). Nutritional effects of peritonitis in continuous ambulatory peritoneal dialysis (CAPD) patient. *J. Am. Diet. Assoc.* 87(1):53-6.

3. Bergstrom, J., Wang, T., & Lindholm, B. (1998). Factors contributing to catabolism in end-stage renal disease patients. *Miner. Electrolyte Metab.* 24(1):92-101.

4. Besarab, A., Bolton, W.K., Browne, J.K., Egrie, J.C., Nissenson, A.R., Okamoto, D.M., Schwab, S.J., & Goodkin, D.A. (1998). The effects of normal as compared with low hematocrit values in patients with cardiac disease who are receiving hemodialysis and epoetin. *N. Engl. J. Med.* 27;339(9):584-90.

5. Chertow, G.M., Bullard, A., & Lazarus, J.M. (1996). Nutrition and the dialysis patient. *Am. J. Nephrol.* 16(1):79-89.

6. Chohan, N. Senior Editor. (1998). Nursing 98 Drug Handbook. Springhouse Corporation, Springhouse, Pennsylvania.

7. Fischbach, F.T. (1995). *A Manual of Laboratory & Diagnostic Tests.* 5th Ed. Philadelphia. J.B. Lippincott Company.

8. Fouque, D. (1997). Causes and interventions for malnutrition in patients undergoing maintenance dialysis. *Blood Purif.* 15(2):112-20.

9. Islam, M.S., Briat, C., Soutif, C., Barnouin, F., & Pollini, J. (1997). More than 17 years of peritoneal dialysis: a case report. *Adv. Perit. Dial.* 13:98-103.

10. Kopple, J.D. (1997). Nutritional status as a predictor of morbidity and mortality in maintenance dialysis patients. *ASAIO J.* 43(3):246-50.

11. Lazarus, J.M. (1993). Nutrition in hemodialysis patients. *Am J. Kidney Dis.* 21(1):99-05.

12. Lindholm, B. & Bergstrom, J. (1992). Nutritional aspects of peritoneal dialysis. *Kidney Int. Suppl.* 38:S165-71.

13. *Manual of Clinical Dietetics.* (1996). 5th Ed. Chicago, IL. The American Di[?] Association.

14. McCusker, F.X. & Teehan, B.P. (1997). Peritoneal dialysis: an evolvin[?] *Semin. Nephrol.* 17(3):226-38.

15. Nursing 98 Books. (1998). *Drug Handbook.* Springhouse, Pennsy[?] Springhouse Corporation.

16.   Prendergast, A. & Fulton, F.L. (1997). *Medical Terminology: A Text/Workbook*. 4th Ed.

17.   Pronsky, Z.M. & Solomon, E. (1998). *Food-Medication Interactions*. 10th Ed. Phoenix, Arizona. Food-Medication Interactions, Publishers and Distributors.

18.   Renal Dietetians Dietetic Practice Group of The American Dietetic Association. (1993). *National Renal Diet: Professional Guide*. Chicago, IL,

19.   Schmicker, R. (1995). Nutritional treatment of hemodialysis and peritoneal dialysis patients. *Artif. Organs*. 19(8):837-41.

20.   Whitney, E.N., Cataldo, C.B., and Rolfes, S.R. (1998). *Understanding Normal and Clinical Nutrition*, 5th Ed. West/Wadsworth.

CASE STUDY #44
GESTATIONAL DIABETES

INTRODUCTION

In this study the diagnosis and treatment of gestational diabetes is examined. Emphasis is placed on the regulation of insulin in conjunction with nutritional management. The student should review the nutritional implications of pregnancy, the medical nutrition therapy for diabetes, and the use of insulin in managing diabetes prior to studying this case.

SKILLS NEEDED

ABBREVIATIONS:

Knowledge of the following abbreviations is required in order to understand this case. You should learn these abbreviations before you begin to read the study.

ABW : adjusted body weight
ACOG : American College of Obstetricians and Gynecologists
A.M. : morning snack around 10:00 A.M.
BMI : body mass index
BS : blood sugar
Ca : calcium
CDE : certified diabetes educator
Cl : chloride
cm : centimeter
DM : diabetes mellitus
Fe : iron
g : gram
GCT : glucose challenge test
GDM : gestational diabetes mellitus
g/dl : grams per deciliter
GTT : glucose tolerance test
HbA1c : glycosylated hemoglobin
Hgb : hemoglobin
Hct : hematocrit
hrs : hours

hs : at bedtime
IVGTT : intravenous glucose tolerance test
K : potassium
kcals : kilocalories
kg : kilogram
mEq : milliequivalent
mg/dl : milligram per deciliter
Na : sodium
NDDG : National Diabetes Data Group
NPH : neutral protamine hagedron insulin
OGTT : oral glucose tolerance test
P.M. : afternoon snack around 3 P.M.
qd : every day
R : regular insulin
RBW : reference body weight
RDA : recommended dietary allowances
Ser Alb : serum albumin
U : units
U/kg : units per kilogram
YOWF : year old white female
μg : microgram

LABORATORY VALUES:

You will need to be able to interpret the nutritional significance of the following laboratory values for this case study: Hgb, Hct, HbA1c, glucose, Ser Alb, K, Na, Cl, transferrin, and urinary ketones (Appendix B).

FORMULAS:

The formulas used in this case study include metric conversions, reference body weight, adjusted body weight, body mass index (Appendix A, Tables A -1, 2, 7 through 9, and 11 through 13), the Harris-Benedict equation, and normal activity factors (Appendix D, Tables D - 1 and 2).

MEDICATIONS:
Become familiar with the following medications before reading the case study. Note the diet-drug interactions, dosages and methods of administration, gastrointestinal tract reactions, etc.
1. Folate; 2. NPH insulin; 3. Regular insulin; 4. Lispro insulin.

Mrs. C is a 31 YOWF who is in the 27th week of her first pregnancy. She is 165.1 cm tall and weighs 81.8 kg. Her weight prior to pregnancy was 72.7 kg. She has not been going to her physician as she should since becoming pregnant, but she was dutiful in making prepregnancy plans. She saw her physician prior to becoming pregnant and followed his advice by taking a prenatal vitamin and 800 μg of folate qd. She was also advised to follow a diet and an exercise plan but Mrs. C has never been able to practice healthy eating habits or exercise regularly. Her family history is positive for diabetes. A grandmother, an aunt, and two cousins were diagnosed with DM during pregnancy. Mrs. C was supposed to see her physician by the 24th week of pregnancy to be screened for GDM but she did not keep her appointment. She is visiting her physician in her 27th week to be screened for GDM.

One hour after consuming 50 g of glucose, Mrs. C's plasma glucose was 165 mg/dL. These results are indicative of GDM and require further testing. Mrs. C agreed to make an appointment to come back to the clinic for an OGTT.
****

QUESTIONS:

1. Convert Mrs. C's height and weight from the metric and determine her RBW (Appendix A, Tables A - 1, 2, and 7 through 10)

2. Mrs. C gained 9.09 kg by her 24th week of pregnancy. What is the recommended weight gain by the 24th week of gestation?

3. Why is it important to take additional folate daily prior to becoming pregnant?

4. According to the Expert Committee on the Diagnosis and Classification of Diabetes Mellitus, Mrs. C has three risk factors for GDM. What are those risk factors? (See reference # 20)

5.   Pregnant women who are a high risk for GDM are recommended to be screened between the _____th and the _____th week of gestation.

Information Box 44 - 1

The screening test for GDM, or the glucose challenge test (GCT), consists of a venous plasma glucose measurement one hour after the consumption of 50 g of oral glucose. This test does not take into account the time of the last meal or the time of day. A value of 140 mg/dl or greater is a positive screening according to the Second, Third, and Fourth International Workshop-Conferences on GDM. A positive screening requires further testing with a GTT. There are two variations of the GTT used for pregnant women: a 75 g OGTT and a 100 g OGTT. There is an ongoing debate in the literature as to which is best, but a positive result with either is acceptable diagnostic criteria for GDM.

The test is performed in the morning after an overnight fast of at least eight hours but no greater than 14 hours. There should be at least three days of unrestricted diet and activity preceding the test. Venous blood is drawn prior to the oral ingestion of the 75 or 100 g glucose load and analyzed for plasma glucose. This is considered time "0." This procedure is repeated after 1, 2, and 3 hours. According to the National Diabetes Data Group (NDDG), in order to obtain a definitive diagnosis of GDM, two or more of the four venous plasma glucose determinations must exceed the levels listed in the table below.

NDDG Diagnostic Criteria for GDM Using the 100 g OGTT[24]

Time in hrs	mg/dl
0	105
1	190
2	165
3	145

According to Carr and Gabbe[7], there are at least three sets of guidelines in the literature used to diagnose GDM, one of which is the NDDG guidelines listed in the table above. The other two are recommended by O'Sullivan and Mahan[26] and Carpenter and Coustan[4]. The American College of Obstetricians and Gynecologists (ACOG) and the Expert Committee on the Diagnosis and Classification of Diabetes Mellitus recommend that two or more of the NDDG values be met or exceeded to make the diagnosis of GDM[7].

A fasting plasma glucose level > 126 mg/dl is sufficient to make a diagnosis of GDM and does not warrant a GTT. In fact, it may be dangerous to administer a GTT under such circumstances. The OGTT is considered the most definitive method of making a positive diagnosis of GDM. Some people, particularly during pregnancy, cannot tolerate an oral load of 100 g of glucose on an empty stomach. An IVGTT is available but does not correlate well with the OGTT.

The results of Mrs. C's OGTT are found in Table 1.

Table 1 Results of OGTT

Time in hrs	Plasma Glucose in mg/dL
0	112
1	235
2	195
3	160

Mrs. C was diagnosed with GDM. Prior to Mrs. C's pregnancy, she consulted with her physician in preparation for pregnancy. At that time the dietitian recommended a diet based on her adjusted body weight. If you were the perinatal dietitian consulting with Mrs. C, you would have access to this information. The answers to the next eight questions will provide that information.
****

QUESTIONS CONTINUED:

6.   Calculate Mrs. C's ABW using her weight prior to pregnancy (Appendix A, Table A - 11).

7.   Calculate Mrs. C's BMI based on her actual body weight prior to pregnancy (Appendix A, Table A - 12).

8.   Based on Mrs. C's BMI, would you classify her as underweight, in her normal weight range, overweight, or obese?

9.   Use the Harris-Benedict equation to calculate Mrs. C's basal metabolic rate. Use her ABW prior to her pregnancy in the formula (Appendix D, Table D - 1).

10. Assume Mrs. C to be lightly active. Choose the appropriate activity factor and determine the daily caloric requirement she would have had prior to pregnancy (Appendix D, Table D - 2).

11. How many kcals do you add for pregnancy with a patient like Mrs. C who is overweight? Explain your answer and if you recommended additional calories, describe what should be the source of those kcals.

12. How much of an additional protein intake is recommended during pregnancy?

13. List the RDAs for the following nutrients prior to and during pregnancy.

Nutrient	RDA for Nonpregnant Women	RDA for Pregnancy
Ca		
P		
Fe		
Vitamin D		
Folate		

****

Mrs. C was hospitalized to get her blood glucose under control. Her initial insulin dose was calculated using the following formula: 0.9 U/kg of body weight. During her stay in the hospital, her blood glucose was closely monitored using a capillary glucose meter. The target plasma glucose levels recommended by the ACOG were used as criteria to determine if additional insulin was needed. Those levels are: fasting, 60 - 90 mg/dL; preprandial, 60 - 105 mg/dL; 1-hour postprandial, not > 130 - 140 mg/dL; 2-hours postprandial, < 120 mg/dL[1]. If her plasma glucose was not in one of those ranges, additional insulin was given using the formula, BS -100 ÷ 20 = U of insulin to be administered. It was also desirable to keep her HgbA1c < 7%. The average BS for Mrs. C 2-hours postprandial was 180 mg/dL. The ultimate goal was to achieve euglycemia to reduce the chances of macrosomia in the newborn.

Mrs. C's initial insulin dosing included R and NPH insulin in three daily injections according to a method adapted from Jovanovic-Peterson and Peterson[17] as reported by Carr and Gabbe[7]. The method consists of administering 4/9 of the total insulin as NPH and 2/9 of the total insulin as R in the morning, 1/6 of the total insulin as R at dinner, and 1/6 of the total insulin as NPH at bedtime.

Mrs. C had a diet order that included three snacks, A.M., P.M., and hs. The dietitian was experimenting with the caloric content of the snacks, trying to find the right combination of kcals and carbohydrate to balance Mrs. C's blood glucose. One of her hs snacks consisted of four graham crackers and 8 oz of skim milk. As part of her new daily routine, Mrs. C is being taught how to measure her urine ketone accumulation in the first voided specimen in the morning.

Several lab tests were also conducted. The results of the pertinent labs are listed in Table 2.

Table 2

TEST	RESULT	NORM	TEST	RESULT	NORM	TEST	RESULT	NORM
Ser Alb	4.0 g/dl	3.9 - 5.0	Hgb	13.0 g/dl	12 - 16	Hct	40%	36 - 48
Glucose	280 mg/dl	65 - 110	Na	140 mEq/L	134 - 145	HbA1c	9.2%	< 6%
K	4.8 mEq/L	3.5 - 5.3	trans-ferrin	296 mg/dL	200 - 400	Cl	103 mEq/L	98 - 106

After reviewing the patient's medical record, the perinatal dietitian, who is also a CDE, interviewed Mrs. C to determine her nutritional intake and daily activity. She determined that Mrs. C was under exercising and overeating, particularly foods high in simple sugars. She calculated a caloric level and protein level for Mrs. C. She advised her to eat more protein-rich foods, complex carbohydrates, and fiber. She discouraged the intake of simple sugars. The dietitian developed a meal plan for Mrs. C that included less than 30 g of CHO for breakfast with none of the CHO coming from juice. The plan included 45% of the kcals from CHO, 25% from protein, and 30% from fat. Mrs. C's urine ketone level that morning was "small," so the RD changed her hs snack to a meat sandwich with skim milk.
****

QUESTIONS CONTINUED:

14. The following terms are in common usage among health professionals working with diabetic pregnant women. Give a brief definition of the terms:

Capillary glucose meter:

Pre- and postprandial:

Euglycemia:

Macrosomia:

15. In question 11, you calculated a caloric level for Mrs. C based on her ABW prior to pregnancy. Since she is now in her 27th week of gestation and has gained 20 lbs, would you recommend the same caloric level or a different one? Explain your answer.

16. Based on the information given, calculate Mrs. C's initial insulin dose and divide it into the three injections. Indicate the appropriate amount of NPH and R to be given in each of the three doses.

17. In the labs reported for Mrs. C, plasma glucose was 280 mg/dL. Determine the additional insulin to be administered as a result of that lab.

18. Explain why the RD recommended a diet 45% CHO, 25% protein, and 30% fat. Show how you could use the results of her labs and her interview to arrive at this diet prescription.

19. Why did the RD recommend a breakfast that is high in protein and has less than 30 g of carbohydrate with no juice?

20. Why is it important to check ketones?

21.   Explain the change in the hs snack.

****

Mrs. C was able to get her blood glucose under control and went home in less than a week.

****

<u>ADDITIONAL OPTIONAL QUESTION</u>:

22.   Plan a day's menu with three meals and three snacks for Mrs. C.

Related References

1.   American College of Obstetricians and Gynecologists (1994): Diabetes and Pregnancy. ACOG Technical Bulletin #200. Washington, DC. *ACOG.*

2.   Bonomo, M., Gandini, M.L., Mastropasqua, A., Begher, C., Valentini, U., Faden, D., & Morabito, A. (1998). Which cutoff level should be used in screening for glucose intolerance in pregnancy? Definition of Screening Methods for Gestational Diabetes Study Group of the Lombardy Section of the Italian Society of Diabetology. *Am. J. Obstet. Gynecol.* 179(1):179-85.

3.   Bobrowski, R.A., Bottoms, S.F., Micallef, J.A., & Dombrowski, M.P. (1996). Is the 50-gram glucose screening test ever diagnostic? *J. Matern. Fetal Med.* 5(6):317-20.

4.   Carpenter, M.W. & Coustan, D.R. (1982). Criteria for screening tests for gestational diabetes. *Am. J. Obstet. Gynecol.* 144:768-73.

5.   Carpenter, M.W., Sady, S.P., Sady, M.A., Haydon, B., Coustan, D.R., & Thompson, P.D. (1990). Effect of maternal weight gain during pregnancy on exercise performance. *J. Appl. Physiol.* 68(3):1173-6.

6.   Carr, D.B. & Gabbe, S. (1998). Gestational Diabetes: Detection, Management, and Implications. *Clinical Diabetes.* 16(1):4-11.

7.   Carr, S.R. (1998). Screening for gestational diabetes mellitus. A perspective in 1998. *Diabetes Care.* Suppl2:B14-8.

8.   Chohan, N. Senior Editor. (1998). Nursing 98 Drug Handbook. Springhouse Corporation, Springhouse, Pennsylvania.

9.   Coustan, D.R., Widness, J.A., Carpenter, M.W., Rotondo, L., & Pratt, D.C. (1987). The "breakfast tolerance test": screening for gestational diabetes with a standardized mixed nutrient meal. *Am. J. Obstet. Gynecol.* 157(5):1113-7.

10. Coustan, D.R. & Carpenter, M.W. (1998). The diagnosis of gestational diabetes. *Diabetes Care.* Suppl2:B5-8.

11. Diamond, T. & Kormas, N. (1997). Possible adverse fetal effect of insulin lispro. *N. Engl. J. Med.* 337:1009.

12. Fischbach, F.T. (1995). *A Manual of Laboratory & Diagnostic Tests.* 5th Ed. Philadelphia. J.B. Lippincott Company.

13. Gabbe, S.G. (1998). The gestational diabetes mellitus conferences. Three are history: focus on the fourth. *Diabetes Care.* Suppl2:B1-2.

14. Gonzalez, M.J., Schmitz, K.J., Matos, M.I., Lopez, D., Rodriguez, J.R., & Gorrin, J.J. (1997). Folate supplementation and neural tube defects: a review of a public health issue. *P.R. Health Sci. J.* 16(4):387-93.

15. Hamaouil, E. & Hamaoui, M. (1998). Nutritional assessment and support during pregnancy. *Gastroenterol. Clin. North Am.* 27(1):89-121.

16. Humalog package insert. Indianapolis, Indiana. Eli Lilly and Company.

17. Jovanovic-Peterson, L. & Peterson, C.M. (1996). Review of gestational diabetes mellitus and low-calorie diet and physical exercise as therapy. *Diabetes Metab. Rev.* 12:287-308.

18. Knoop, R.H., Magee, M.S., Raisys, V., Benedetti, T., & Bonet, B. (1991). Hypocaloric diets and ketogenesis in the management of obese gestational diabetic women. *J. Am. Coll. Nutr.* 10(6):649-67.

19. Kuhl, C. (1998). Etiology and pathogenesis of gestational diabetes. *Diabetes Care.* Suppl2:B19-26.

20. Lloyd, K. Editor. (1998). *A Core Curriculum for Diabetic Educators.* American Association of Diabetes Educators. Chicago, IL. Port City Press, Inc.

21. Locksmith, G.J. & Duff, P. (1998). Preventing neural tube defects: the importance of periconceptional folic acid supplements. *Obstet. Gynecol.* 91(6):1027-34.

22. Magee, M.S., Knoop, R.H., & Benedetti, T.J. (1990). Metabolic effects of a 1200 kcal diet in obese pregnant women with gestational diabetes. *Diabetes.* 39:234-40.

23. *Manual of Clinical Dietetics.* (1996). 5th Ed. Chicago, IL. The American Dietetic Association.

24. National Diabetes Data Group. (1979). Classification and diagnosis of diabetes mellitus and other catagories of glucose intolerance. *Diabetes.* 28:1039-57.

25. Nursing 98 Books. (1998). *Drug Handbook.* Springhouse, Pennsylvania. Springhouse Corporation.

26. O'Sullivan, J.B. & Mahan, C.M. (1964). Criteria for the oral glucose tolerance test in pregnancy. *Diabetes.* 13:278-85.

27. Prendergast, A. & Fulton, F.L. (1997). *Medical Terminology: A Text/Workbook.* 4th Ed.

28. Pronsky, Z.M. & Solomon, E. (1998). *Food-Medication Interactions.* 10th Ed. Phoenix, Arizona. Food-Medication Interactions, Publishers and Distributors.

29. Robinson. S., Godfrey, K., Denne, J., & Cox, V. (1998). The determinants of iron status in early pregnancy. *Br. J. Nutr.* 79(3):249-55.

30. Sacks, D.A., Abu-Fadil, S., Greenspoon, J.S., & Fotheringham, N. (1989). Do the current standards for glucose tolerance testing represent a valid conversion of O'Sullivan's original criteria? *Am. J. Obstet. Gynecol.* 161:638-41.

31. Sullivan, B.A., Henderson, S.T., & Davis, J.M. (1998). Gestational Diabetes. *J. Am. Pharm. Assoc.* 38(3):364-73.

32. Tallarigo, L., Giampietro, O., Penno, G., Miccose, R., Gregori, G., & Navalesi, R. (1986). Relation of glucose tolerance to complications of pregnancy in nondiabetic women. *N. Engl. J. Med.* 315:989-92.

33.   Weiner, C.P. (1988). Effect of varying degrees of normal glucose metabolism on maternal and perinatal outcome. *Am. J. Obstet. Gynecol.* 159:862-70.

34.   Weiss, P.A., Haeusler, M., Kainer, F. Purstner, P., & Hass, J. (1998). Toward universal criteria for gestational diabetes: relationships between seventy-five and one hundred gram glucose loads and between capillary and venous glucose concentrations. *Am. J. Obstet Gynecol.* 174(4):830-5.

35.   Whitney, E.N., Cataldo, C.B., and Rolfes, S.R. (1998). *Understanding Normal and Clinical Nutrition*, 5th Ed. West/Wadsworth.

APPENDIX A

Commonly Used Formulas and Conversion Factors

Table A - 1
Weight Conversions:

1 kilogram (kg) = 1,000 grams (g)	1 ounce (oz) = 28 grams
1 gram (g) = 1,000 milligrams (mg)	16 ounces = 1 pound (lb)
1 milligram = 1,000 micrograms (μg)	1 pound = 454 grams
1 microgram = 1,000 nanograms (ng)	1 kilogram = 2.2 pounds
1 nanogram = 1,000 picograms (pg)	1 teaspoon (tsp) = ~ 5 grams
	3 teaspoons = ~ 1 tablespoon (tbs)

Table A - 2
Volume conversions:

1 liter (L) = 1,000 milliliters (mL or ml)	1 cup (c) = 8 fluid ounces
1 liter = 1.06 quarts (qt)	1 cup = ~ 240 milliliters
1 quart = 0.95 liters	4 cups = 1 quart
1 milliliter = 1 cubic centimeter (cc)	1 teaspoon (tsp) = ~ 5 milliliters
30 milliliters = 1 fluid ounce	1 tablespoon (tbs) = ~ 15 milliliters
32 ounces = 1 quart	2 tablespoons = ~ 1 ounce

Table A - 3
Length Conversions

Table A - 4
Temperature Conversions

1 inch = 2.54 centimeters (cm) 1 foot (ft) = 30.48 centimeters 1 meter (m) = 100 centimeters 1 meter = 39.37 inches 1 meter = 3.28 feet	To convert Celsius (Centigrade) to Fahrenheit: $(9/5 \times t_c) + 32 = t_F$ To convert Fahrenheit to Celsius: $5/9 (t_F - 32) = t_c$

Table A - 5
Milligrams and Milliequivalents:

To convert milliequivalents (mEq) to milligrams (mg), multiply mEq by the element's atomic weight and divide by the valence: mEq X atomic weight/valence = mg
To convert milligrams to milliequivalents, divide milligrams by the element's atomic weight and multiply by the valence: mg/atomic weight X valence = mg

Example: To convert 2000 mg of Na to mEq: 2000/23 X 1 = 86.96 mEq	**Example:** To convert 80 mEq of K to mg: 80 X 39/1 = 3120 mg

Table A - 6
Atomic Weights and Valences

	Atomic Weight	Valence		Atomic Weight	Valence
Na	23	1	Mg	24.3	2
K	39	1	P	31	2
Cl	35.4	1	Ca	40	2

Table A - 7
Reference Body Weight (RBW):

The Hamwi formula[1] is a quick method commonly used by dietitians to determine someone's "ideal body weight." The term "ideal" has been questioned because, with changing weight standards and the possible significant variation between individuals, the "ideal weight" is difficult to define. *The Dietary Guidelines for Americans*[2], revised in 1995, suggested a new weight range for all ages and both genders. The lower end of the range is intended for less muscular individuals and the higher end for those who are more muscular. The new dietary guidelines are used throughout this book with the "midpoint" of the range being referred to as the "reference body weight" (RBW). The RBW can be adjusted from the midpoint to anywhere in the range depending on the muscularity of the individual being assessed. The weight ranges recommended by *The Dietary Guidelines for Americans*[2], along with the midpoints, can be found in **Table A - 9**. If you prefer, you may continue to use the IBW as described by Hamwi in the next table.

Table A - 8
Ideal Body Weight (IBW):

If the Hamwi[1] method is preferred, it recommends the following procedures:
For males, it allows 106 pounds for the first 5 feet plus 6 pounds for every additional inch over 5 feet.
For females, 100 pounds is allowed for the first 5 feet plus 5 pounds for each additional inch.
Up to 10% can be added for a large frame and up to 10% can be subtracted for a small frame. For a comparison of the IBW and the new dietary guidelines for Americans, see **Table A - 9**.

Example: Calculate the IBW of a 5'11" male with a very large frame. For the first 5' = 106 lbs For the 11" over 5' add 11 X 6 = 66 IBW = 106 + 66 = 172 ± 10% For a very large frame: 172 X 10% = 17.2 or 17 IBW = 172 + 17 = 189 lbs	**Example:** Calculate the IBW of a 5'4" female with an average frame. For the first 5' = 100 lbs For the 4" over 5' add 4 X 5 = 20 IBW = 100 + 20 = 120 ± 10%

Table A - 9
1995 Weight Guidelines Compared to IBW:

1995 Guidelines[2] for Adults of All Ages			IBW Calculater by the Hamwi[1] Method			
Height[a]	Midpoint	Range	Male IBW ± 10%		Female IBW	± 10%
4'10"	105	91 - 119				
4'11"	109	94 - 124				
5'0"	112	97 - 128	106	95 - 117	100	90 - 110
5'1"	116	101 - 132	112	102 - 123	105	94 - 116
5'2"	120	103 - 137	118	106 - 130	110	99 - 121
5'3"	124	107 - 141	124	112 - 136	115	103 - 127
5'4"	128	111 - 146	130	117 - 143	120	108 - 132
5'5"	132	114 - 150	136	122 - 150	125	113 - 138
5'6"	136	118 - 155	142	128 - 156	130	117 - 143
5'7"	140	121 - 160	148	135 - 163	135	121 - 149
5'8"	144	125 - 164	154	139 - 169	140	126 - 154
5'9"	149	129 - 169	160	144 - 176	145	130 - 160
5'10"	153	132 - 174	166	149 - 183	150	135 - 165
5'11"	157	136 - 179	172	155 - 189	155	139 - 171
6'0"	162	140 - 184	178	160 - 196	160	144 - 176
6'1"	166	144 - 189	184	166 - 202	165	148 - 182
6'2"	171	148 - 195	190	171 - 209	170	153 - 187
6'3"	176	152 - 200	196	176 - 216	175	157 - 193
6'4"	180	156 - 205	202	182 - 222	180	162 - 198
6'5"	185	160 - 211	208	187 - 229	185	166 - 204
6'6"	190	164 - 216	214	193 - 235	190	171 - 209

Table A - 10
Percentages:

%RBW and %UBW	% of Weight Loss
$\% \text{ RBW } = \dfrac{\text{Actual Weight}}{\text{Reference Weight}} \times 100$	Original Weight - Final Weight = Loss of Weight
$\% \text{ UBW } = \dfrac{\text{Actual Weight}}{\text{Usual Weight}} \times 100$	$\dfrac{\text{Loss of Weight}}{\text{Original Weight}} = \% \text{ of Weight Loss}$

Table A - 11
Adjusted Body Weight[3] (ABW)

Adjusted Body Weight = [(actual body weight - IBW or RBW) X 0.25[a]] + IBW or RBW

[a] This is based on data that indicates that only 25% of body adipose tissue is metabolically active. When this formula was developed, it was based on the then popular ideal body weight. It was assumed that this formula allowed for a uniform estimation of the caloric needs of the obese. If the assumption is true for IBW, it should be true for RBW also. However, there are some who do not like this assumption and do not use this formula.

Table A - 12
Body Mass Index[4]:

Metric	English[5]
$BMI = \dfrac{\text{weight in kg}}{\text{height in m}^2}$	$BMI = \dfrac{\text{weight in lbs}}{\text{height in inches}^2}$

Table A - 13
Relationship between BMI and obesity:

In 1989, Rowland developed a nomogram for computing body mass index[6]. The nomogram gives the following classifications for BMI:

	Men	Women
Underweight	< 20.7	< 19.1
Acceptable weight	20.7 - 26.4	19.1 - 25.8
Marginal overweight	26.4 - 27.8	25.8 - 27.3
Overweight	27.8 - 31.1	27.3 - 32.2
Severe overweight	31.1 - 45.4	32.3 - 44.8
Morbid obesity	> 45.4	> 44.8

In 1995, in a report produced by the Dietary Guidelines Advisory Committee[2], emphasis was given to the importance of characterizing the upper level of a healthy weight based on a pathologic sequelae rather than an arbitrary definition. The report asked the question: Should the cutoff be based on morbidity or mortality? Their answer was that mortality is the most significant and reliable consequence of disease and should therefore be the determining factor. This creates a problem in interpreting the classifications of BMI. For example, in their classifications, a BMI over 25 is considered obese. Research has shown that as BMI increases from 25, mortality significantly increases[7, 8, 9]. However, research has also shown that the incidence of diabetes begins to increase with a BMI below 25 [10, 11]. A BMI between 20 and 25 is considered "normal," but, as the number approaches 25, it could mean a risk for diabetes for some individuals.

Table A - 14
Midarm Muscle Circumference (MAMC)[12, 13]:

MAMC (cm) = midarm circumference (cm) - [3.14* X triceps skinfolds (mm)]

* This factor converts the fatfold measurement to a circumference measurement and millimeters to centimeters.

Table A - 15
Total Lymphocyte Count (TLC):

TLC (mm^3) = WBC mm^3 x % Lymphocytes

Table A - 16
Rule of Nines[14]:

The "Rule of Nines" is used to obtain an estimate of the body surface area burned. With this method, each of the following body areas is assumed to cover the indicated percent of the body surface area.

Each arm	= 9%	The posterior trunk	= 18%
Each leg	= 18%	The head	= 9%
The anterior trunk	= 18%	The perineum	= 1%

Table A - 17
Nitrogen Balance[15]:

The determination of nitrogen balance requires knowing the 24-hour nitrogen intake and output. To estimate nitrogen from dietary protein, use either of these formulas:

Nitrogen Intake = $\underline{\text{Protein Intake (g)}}$ or   Protein Intake X 16% = Nitrogen Intake
                6.25

Nitrogen output per day equals the measured UUN plus a factor of 4 grams to account for nitrogen losses through the lungs, hair, skin, and nails as well as non-urea nitrogen losses in the urine.

Nitrogen output = UUN + 4g

Nitrogen Balance = $\underline{\text{Protein Intake (g)}}$ - (UUN + 4g)   or   Protein Intake X 16% = Nitrogen Intake
                    6.25

REFERENCES

1.   Miller, M.A. (1985). A calculated method for determination of ideal body weight. *Nutri Support Services.* pp. 31-33.

2.   *Report of the Dietary Guidelines Advisory Committee on the Dietary Guidelines for Americans.* (1995). Washington, D.C., Government Printing Office.

3.   Karkeck, J.M. (1984). Adjustment for Obesity, *Am. Diet. Assoc. Renal Practice Group Newsletter*, Winter Issue.

4.   Garrow, J.S. & Webster, J. (1985). Quetelet's index (W/H^2) as a measure of fatness. *Int. J. Obesity.* 9:147-153.

5.   Stensland, S.H. & Margolis, S. (1990). Simplifying the calculation of body mass index for quick reference. *J. Am. Diet. Assoc.* 90:856.

6.   Rowland, M.I. (1989). A nomogram for computing body mass index. *Dietetic Currents.* 16:8-9. Columbus, Ohio. Ross Laboratories.

7.   Lee, I.M. & Paffenbarger, R.S., Jr. (1992). Change in body weight and longevity. *JAMA.* 268:2045-2049.

8.   Willett, W.C., Manson, J.E., Stampfer, M.F.,Colditz, G.A., Rosner, B, Speizer, F.E., & Hennekens, C.H. (1995) Weight, weight change, and coronary heart disease in women. Risk within the 'normal' weight range. *JAMA.* 273(6):461-465.

9.   Rimm, E.B., Stampfer, M.J., Giovannucci, E., Ascherio, A., Spiegelman, D., Colditz, G.A., & Willett, W.C. (1995). Body size and fat distribution as predictors of coronary heart disease among middle-aged and older US men. *Am. J. Epidem.* 141:1117-1127.

10.   Chan, J.M., Rimm, E.B., Colditz, G.A., Stampfer, M.J., & Willett, W.C. Obesity, fat distribution, and weight gain as risk factors for clinical diabetes in men. Diabetes Care. 17(9):961-969.

11.   Colditz, G.A., Willett, W.C., Rotnitzky, A., & Manson, J.E. (1995). Weight gain as a risk factor for clinical diabetes mellitus in women. *Ann. Inter. Med.* 122:548-549.

12.   Whitney, E.N., Cataldo, C.B. & Rolfes, S.R. (1998). *Understanding Normal & Clinical Nutrition.* 5th Ed. West/Wadsworth Publishing Company. pg. E23 - E27.

13.   Lee, R.D. & Nieman, D.C. (1993). *Nutritional Assessment.* Madison, Wisconsin. Brown & Benchmark, Publishers. pg. 168 - 174.

14.   Rakel, R.E. & Conn, H.F. Eds. (1978) The Rule of Nines. Family Practice, 2nd Ed. Philadelphia, W.B. Saunders Company. pg. 536-537.

15.   Lee, R.D. & Nieman, D.C. (1993). *Nutritional Assessment.* Madison, Wisconsin. Brown & Benchmark, Publishers. pg. 226.

APPENDIX B

BLOOD, PLASMA, OR SERUM LABORATORY VALUES[1]

REFERENCE RANGES

Determination	Conventional Units	SI Units
Albumin	3.9 - 5.0 g/dl	39 - 50 g/L
Alk Phos (ALP)	17 - 142 U/L	17 - 142 U/L
ALT (SGPT)	7 - 56 U/L	7 - 56 U/L
Ammonia	9 - 33 µg/dl	9 - 33 µmol/L
Amylase	25 - 125 U/L	25 - 125 U/L
AST (SGOT)	5 - 40 U/L	5 - 40 U/L
Bilirubin	Total 0.2 - 1.0 mg/dl	3.4 - 17.1 μmol/L
	Direct 0.0 - 0.2 mg/dl	0 - 3.4 μmol/L
BUN	7 - 18 mg/dl	2.5 - 6.4 mmol/L
Ca	Total 8.6 - 10.0 mg/dl	2.15 - 2.50 mmol/L
	Ionized 4.65 - 5.28 mg/dl	1.16 - 1.32 mmol/L
Cholesterol	140 - 199 mg/dl	<5.18 mmol/L
CPK		
Men	38 - 174 IU/L	38 - 174 U/L
Women	96 - 140 IU/L	96 - 140 U/L
	Isoenzymes	
	CPK-BB (CPK$_1$) 0%	
	CPK-MB (CPK$_2$) 0 - 4 %	
	CPK-MM (CPK$_3$) 96 - 100 %	
Cl	98 - 106 mEq/L	98 - 106 mmol/L
Cr	0.6 - 1.3 mg/dl	62 - 115 μmol/L
Fe	75 - 175 μg/dl	13.4 - 31.3 μmol/L
GGT		
Men	5 - 85 U/L	5 - 85 U/L
Women	5 - 55 U/L	5 - 55 U/L
Glucose	fasting 65 - 110 mg/dl	3.5 - 6.1 mmol/L
GTT	30 min 110 - 170 mg/dl	6.1 - 9.4 mmol/L
	60 min 120 - 170 mg/dl	6.7 - 9.4 mmol/L
	120 min 70 - 120 mg/dl	3.9 - 6.7 mmol/L
	3 hr 70 - 120 mg/dl	3.9 - 6.7 mmol/L
HCO$_3$	24 - 28 mEq/l	24 - 28 mmol
Hct	M 42 - 52%	
	F 36 - 48%	
HDL	M 37 - 70 mg/dl	.95 - 1.8 mmol/L
	F 40 - 85 mg/dl	1.0 - 2.2 mmol/L
HbA1c		
Nondiabetic	5.5 - 8.5%	
Diabetic:		
Good Control	7.5 - 11.4%	
Moderate Control	11.5 - 15%	
Poor Control	>15%	

Hgb	M 14 - 17.4 g/dl	2.17 - 2.70 nmol/L
	F 12 - 16 g/dl	1.86 - 2.48 nmol/L
K	3.5 - 5.3 mEq/L	3.5 - 5.3 mmol/L
	Panic value <2.5 - >7.0	2.5 - 7.0 mmol/L
LDH	313 - 618 U/L	313 - 618 U/L
	Isoenzymes	
	LD_1 17 - 27%	
	LD_2 29 - 39%	
	LD_3 19 - 27%	
	LD_4 8 - 16%	
	LD_5 6 - 16%	
LDL	<130 mg/dl	<3.4mmol/L
Lymphocytes	20 - 40% of total leukocytes	Relative value = 1000 - 4000 mm^3
MCH	26 - 34 pg/cell	.40 - .53 fmol/cell
MCV	82 - 98 μm^3	82 - 98 fL
Mg	1.3 - 2.1 mEq/L	0.65 - 1.0 5 mmol/L
Na	135 - 145 mEq/L	135 - 145 mmol/L
	Panic value <120 - >155	120 - 155 mmol/L
Osmolality	275 - 295 mOsm/kg	275 - 295 mmol/kg
P	2.5 - 4.5 mg/dl	0.87 - 1.45 mmol/L
$PaCO_2$	35 - 45 mm Hg	4.66 - 5.99 kPa
PaO_2	80 - 90 mm Hg	10.64 - 11.97 kPa
pH	7.35 - 7.45	
Platelet Count	140,000 - 400,000/mm^3	140,000 - 400,000 $x10^6$/L
Prealbumin	10 - 40 mg/dl	
PT	10 - 13 secs.	
Sed Rate	Westergren method	
	M 0 - 15 mm/hr	
	F 0 - 20 mm/hr	
TLC	>1500/mm^3	
Triglycerides	40 - 160 mg/d l	0.45 - 1.8 mmol/L
Transferrin	200 - 400 mg/dl	
Uric Acid	F 2.6 - 6.0 mg/dl	0.15 - 0.35 μmol/L
	M 3.5 - 7.2 mg/dl	0.21 - 0.42 μmol/L
WBC	5 - 10 x 10^3/μl	5 10 x 10^9/L

PEDIATRICS

Albumin	1 - 3 yrs	5.9 - 7.0 g/dl
BUN	child	5 - 18 mg/dl
Glucose	7 d - 6 yr	74 - 127 mg/dl
Hemoglobin	1 - 6 yrs	9.5 - 14.1 g/dl
K	1 - 18 yrs	3.4 - 4.7 mEq/l

URINALYSIS:

Bili	Neg
Blood	Neg
Glucose	Neg
Ketones	Neg
pH	4.5 - 8
Protein	Neg
SG	1.015 - 1.025

[1] Laboratory reference values will vary from hospital to hospital depending on the type instrumentation used in analysis, reference standards used to calibrate the instrumentation, and other factors. Each hospital produces their own list of "norms." The items in this list are from F. Fischbach's *Laboratory & Diagnostic Tests,* 5th Ed., 1996, Lippincott-Raven. Furthermore, this list is not conclusive but contains only those values that are necessary for answering the case studies in this book. Most of the labs in this book are reported in conventional units. The corresponding SI Units are listed here for your convenience.

ABBREVIATIONS

AA	Amino acid
ABGs	Arterial blood gasses
a.c.	Before meals
ACOG	American College of Obstetricians and Gynecologists
ADD	Attention deficit disorder
ADHD	Attention deficit hyperactive disorder
AIDS	Acquired immune deficiency syndrome
Alk Phos (ALP)	Alkaline phosphatase
ALT (formally SGPT)	Alanine aminotransferase
AM	Morning, A.M. snack
amp	Ampule
AODM	Adult onset diabetes mellitus
ARC	AIDS Related Complex
ASA	Acetylsalicylic acid (aspirin)
ASAP	As soon as possible
AST (formally SGOT)	Aspartate aminotransferase
A-V	Arteriovenous shunt or atrioventricular
BCAA	Branched chain amino acids
BE	Barium enema, below the elbow
BEE	Basal energy expenditure
BIA	Bioelectrical impedance
b.i.d.	Twice a day
bili	Bilirubin
BM	Black male
BMR	Basal metabolic rate
BMI	Body mass index
BP	Blood pressure
BPD	Broncopulmonary displesia
BR	Bed rest
BRB	Bright red blood
BS	Bowel sounds, blood sugar, breath sounds
BSA	Body surface area
BUN	Blood urea nitrogen
BUT	Biopsy urease test
C	Centigrade
c	Cup
c̲	With
CA	Carcinoma
Ca	Calcium
CABG	Coronary artery by-pass graft
CAPD	Continuous ambulatory peritoneal dialysis
Cardiac cath	Cardiac catheterization
cath	Catheterization
CC	Chief complaint

cc	Cubic centimeter
cc/h	Cubic centimeter per hour
CCU	Coronary care unit
CDE	Certified diabetes educator
CF	Cystic fibrosis
CHD	Coronary heart disease
CHF	Congestive heart failure
CHI	Closed head injury
CHO	Carbohydrate
chol	Cholesterol
Cl	Chlorine
cl liqs	Clear liquids
cm	Centimeter
C/O	Complains of
CPK	Creatine phosphokinase
CPK_{1-3}	Creatine phosphokinase isoenzymes
Cr	Creatinine
CxR	Chest X Ray
CVR	Cerebrovascular accident
d	Day
DBW	Desirable body weight
D/C	Discontinue
DKA	Diabetic ketoacidosis
dl	Deciliter
DM	Diabetes mellitus
D_5NS	Dextrose, 5% in normal saline
DT	Delirium tremens
D_5W	Dextrose, 5% in water
$D_{10}W$	Dextrose, 10% in water
$D_{50}W$	Dextrose, 50% in water
DVT	Deep vein thrombosis
Dx	Diagnosis
EGD	Esophagogastroduodenoscopy
EKG	Electrocardiogram
Elisa	Enzyme linked immunosorbent assay
ER	Emergency room
ESRD	End stage renal disease
ETOH	Ethanol or alcohol
Exploratory Lap	Exploratory laparotomy
F	Fahrenheit
Fe	Iron
FF	Fource fluids
f	femto (10^{-15})
FBS	Fasting blood sugar
FH	Family history
fL	femtoliter
Fx	Fracture

g	Gram
GB	Gallbladder
GCT	Glucose challenge test
g/dl	grams per deciliter
GDM	Gestational diabetes mellitus
GERD	Gastroesophageal reflux disease
GGTP	Gama glutamyl transpeptidase
GI	Gastrointestinal
glu	Glucose
G-tube	Same as PEG
GTT	Glucose tolerance test
h	Hour
HA	Head ache
HbA1c	Glycosylated hemoglobin
HBC	High branched chain
HCl	Hydrochloric acid
HCO_3	Bicarbonate ion
Hct	Hematocrit
H & H	Hematocrit and hemoglobin
HD	Hemodialysis
HDL	High density lipoprotein
Hg	Mercury
Hgb	Hemoglobin
HPRL	Prolactin
HIV	Human immunodeficiency virus
HN	High nitrogen
h.s.	Hour of sleep or evening
Ht	Height
HTN	Hypertension
Hx	History
IBD	Inflammatory bowel disease
IBW	Ideal body weight
ICP	Intracranial pressure
ICU	Intensive care unit
IDDM	Insulin dependent diabetes mellitus
IgG	Immunoglobin g
IHDP	Infant Health and Development Program
I.V.	Intravenous
IVGTT	Intravenous glucose tolerance test
IM	Intramuscular
in	Inch
IU	International units
JODM	Juvenile onset diabetes
K	Potassium
k	kilo
kcals	Kilocalories

kg	Kilogram
KUB	Kidney, ureter, and bladder (X-ray)
L	Liter
LAP	Laparotomy
LBM	Lean body mass
lbs	Pounds
LD_{1-5}	Isoenzymes of LDH
LDH	Lactic dehydrogenase
LDL	Low density lipoprotein
LES	Lower esophageal sphincter
LLQ	Left lower quadrant
LUQ	Left upper quadrant
lymph	Lymphocytes
lytes	Electrolytes
MAC	Midarm circumference
MAMC	Midarm muscle circumference
mcg	microgram
MCH	Mean corpuscular hemoglobin
mci	millicuries
MCV	Mean corpuscular volume
MD	Medical doctor
MDI	Multiple daily injections
mech	Mechanical
mEq	Milliequivalent
Mg	Magnesium
mg	Milligram
mg/dl	Milligram per deciliter
$MgSO_4$	Magnesium sulfate
MH	Medical history
MI	Myocardial infarction
milliIU/L	Milliinternational units per iter
min	Minute
mm	Millimeter
mm^3	Cubic millimeter
mmol	Millimole
MNT	Medical nutrition therapy
mol	mole
MOM	Milk of magnesia
mos	Months
mOsm	Milliosmole
MS	Morphine sulfate
MTE	Mixture of trace elements
MVA	Motor vehicle accident
MVI	Multiple vitamin infusion
N	Nitrogen
Na	Sodium
NDDG	National diabetes Data Group
neg	Negative

ng	Nanogram
N/G	Nasogastric
NH	Nutritional history
NH$_3$	Ammonia
NIDDM	Noninsulin dependent diabetes mellitus
NPH	Neutral protamine Hagedorn insulin
NPO	Nothing by mouth
NS	Normal saline
NSICU	Neurosurgical intensive care unit
NST	Nutrition support team
NTG	Nitroglycerin
N/V	Nausea and vomiting
O$_2$	Oxygen
OGTT	Oral glucose tolerance test
OR	Operating room
oz	Ounce
p̄	After
P	Phosphorous
Pa	Pressure
PaCO$_2$	Partial pressure of carbon dioxide
PaO$_2$	Partial pressure of oxygen
PC	Packed cells
p.c.	After meals
PD	Peritoneal dialysis
PED	Percutaneous endoscopic duodenostomy
PEG	Percutaneous endoscopic gastrostomy
pg	Picogram
pH	The negative logarithm of the hydrogen ion concentration
PICU	Pediatric intensive care unit
PM	Afternoon, P.M. snack
p.o.	By mouth
Post-op	Post-operative
prn	As needed
prot	Protein
PT	Prothrombin time, Physical therapy
pt	Patient
PUD	Peptic ulcer disease
PVC	Premature ventricular contraction
PVD	Peripheral vascular disease
q AM	Every morning
qd	Every day
qid	Four times a day
q4h	Every 4 hours
q6h	Every 6 hours
qod	Every other day
qt	Quart
R	Regular

RBW	Reference body weight
RD	Registered dietitian
RDA	Recommended dietary allowances
RLQ	Right lower quadrant
Rm	Room
RN	Registered nurse
R/O	Rule out
RUQ	Right upper quadrant
Rx	Prescription
sat	Saturated
SBO	Small bowel obstruction
SBR	Strict bed rest
SBS	Short bowel syndrome
SC	Subcutaneous
sec	Second
Sed Rate	Sedimentation rate
Ser alb	Serum albumin
SG	Specific gravity
SGOT	Serum glutamic-oxaloacetic transaminase
SGPT	Serum glutamic-pyruvic transaminase
SH	Social history
SMAC 20	Group of 20 blood tests
SOB	Short of breath
S/P	Status post
T	Tablespoon, temperature
tab	Tablet
TF	Tube feeding
t.i.d.	Three times a day
TLC	Total lymphocyte count, tender loving care
tot	Total
TPN	Total parenteral nutrition
trach	Tracheostomy
trig	Triglycerides
TSF	Triceps skin fold
TSH	Thyroid-stimulating hormone
tsp	Teaspoon
T_3	Free triiodothyronine
T_4	Free thyroxine
U	Unit
UBW	Usual body weight
U/kg	Units per kilogram
UGI	Upper gastrointestinal
U/L	Units per liter
UTI	Urinary tract infection
UUN	Urinary urea nitrogen
VLCD	Very low calorie diet
VLBW	Very low birth weight

WBC	White blood cell count
Wk	Week
WNL	Within normal limits
Wt	Weight
x	Times
x1d	Times one day
x3	Times three
x3d	Times three days
YO	Year old
YOF	Year ole female
YOBF	Year old black female
YOBM	Year black male
YOW	Year old white
YOWF	Year old white female
YOWM	Year old white male

SYMBOLS

@	At
°	Degree
1h	One hour
1st°	First degree
2nd°	Second degree
2x	Two times
2xd	Two times a day
3rd°	Third degree
3x	Three times
2°	Secondary
μ	Micro
μg	Microgram
μm^3	Cubic microns
μmol	Micromole
¼NS	One fourth strength normal saline
♂	Male
♀	Female
↑	Increase
↓	Decrease
∅	None
#	Number
Ⓛ	Left
Ⓡ	Right

NUMBERS

i	One
ii	Two
iii	Three
iv	Four
v	Five
vi	Six
vii	Seven
viii	Eight
ix	Nine
x	Ten

ENERGY DETERMINATIONS: ACTIVITY and STRESS FACTORS[1]

Table D - 1
BEE (Harris-Benedict equation[1]):

Women BEE = 655 + [9.6 x wt[a]] + [1.7 x ht[b]] - [4.7 x age[c]]	Men BEE = 66 + [13.7 x wt[a]] + [5 x ht[b]] - [6.8 x age[c]]
[a]wt is in kg. [b]ht is in cm. [c]age is in yrs	

Table D - 2
Activity Factors for Normal Healthy Individuals:

To determine the total calories required in a day for the ambulatory healthy person, the following activity factors are recommended in the 10th Ed. of the RDAs[2]: Multiply the BMR times the activity factor and add the result to the BMR.

For a sedentary, very inactive lifestyle, add 30% of the BMR (or BMR x 1.3).

For a lightly active lifestyle, add 50% for women (or 1.5 x BMR) and 60% for men (or 1.6 x BMR)

For a moderately active lifestyle, add 60% for women and 70% for men.

For a high level of intense activity, add 90% for women and 110% for men (or 2.1 x BMR).

The figure used in this range depends on the intensity of the activity.

Table D - 3
Activity Factors for Starvation:

According to Weinsier, Heimburger, and Butterworth, the caloric goals in refeeding the hypometabolic, starved patient are as follows[3]:

Days	Calories
1,2	BEE x 0.8*
3,4	BEE x 1.0
4 to 6	BEE x 1.2 to 1.5
6 and after	BEE x 2.0 if weight gain is desired.

*The earlier reference on refeeding the starved patient recommended 60 to 80% of the BEE[4].

Table D - 4
Determining Energy and Protein Needs for a Fever[5]:

BEE increases 13% per degree C or 7% per degree F. To calculate, complete the following steps:

1. Determine BEE
2. Normal temperature = 98.6 F. Elevated temp - normal temp = # of degrees F elevated
3. # of degrees F elevated x 7% = % BEE is to be increased
4. BEE x % calculated in 3. = amount BEE needs to be increased
5. BEE + results in 4. = Total kcals Needed

Example: Assume you calculated a BEE of 1500 kcals for someone who has a fever of 101° F.

1. BEE = 1500 kcals
2. 101 - 98.6 = 2.4° elevated
3. 2.4 x 7% = 16.8%
4. 1500 x 16.8% = 252 kcals
5. BEE + Needs increase = 1500 + 252 = 1752 total kcals

Table D - 5
Stress Factors for Energy and Protein Requirements:

Energy	Protein
To determine the total calories required in a day for the hospitalized or bedridden person, the following stress factors are generally used. Multiply the BEE by the activity factor and add the result to the BEE.	The recommended protein intake per the RDAs[2] for the average person is 8g/kg of RBW. For the athlete, the requirement increases to 1.2 - 1.5[2] or 1.4 - 1.8[6], depending on what reference you use.
20% of the BEE for patients on total bed rest 30% for recovering patients who are ambulatory These are old recommendations but are still used by many practitioners[7]. 5% for ICU patients[8, 9, 10].	Protein intake depends on the amount of weight lost during the sickness, if any, and on the amount of wound healing taking place, blood protein levels, etc.
For liver disease: Energy[11]: 35 to 45 kcal/kg actual body weight (dry weight).	Protein:Cirrhosis without encephalopathy: 1.0 - 1.5 g/kg of dry-body wt[11]. With encephalopathy the current thinking is not so much restriction as prevention of excess. One recommendation is 40 to 60 g of protein per day[11]. O'Keefe et al has stated that if less than 60 g is given per day, protein depletion will take place[12].
BCAA and liver disease: Research surrounding HBCAA is controversial. Cerra et al reported benefit from HBCAA when between 80 and 120 g/d[13] were administered via TPN. A meta analysis showed only slight but significant improvement in hepatic encephalopathy. The effects on mortality seem to be inconsistent[14]. The American Society for Parenteral and Enteral Nutrition clinical guidelines recommend that BCAA supplemented diets not be used routinely for critically ill patients because proven effects on clinical outcomes are lacking[15].	
For chronic renal failure: Energy = 35 - 40 kcals/kg for renal insufficiency 35 - 40 kcals/kg for hemodialysis 25 - 35 kcals/kg for peritoneal dialysis[16].	Protein = 0.6 - 0.8 g/kg for renal insufficiency 0.8 - 1.0 g/kg for nephrotic syndrome 1.1 - 1.4 g/kg for hemodialysis 1.2 to 1.5 for peritoneal dialysis[16].
For kidney transplant[17]: Energy intake in accordance with patient's need to gain or lose weight. Remember that immunosuppressants can induce weight gain. .	Protein intake between 1.2 and 1.5 g/kg of body weight according to patient's needs to restore lab values to normal. Remember that immunosuppressants can induce protein catabolism.
For cancer[18]: Energy intake will vary depending on a number of assessment factors but will usually be around 1.5 x the BEE.	Protein intake could be 1.5 - 2.0 g/kg of ideal body weight[18].

For surgery, sepsis or trauma:	Protein: The protein intake will usually be close to the stress factor x kg of body weight up to about 1.5 g/kg. Between 1.2 and 1.5 g of protein per kg of body weight will meet the needs of most stressed patients. Burns may require more depending on oozing of body fluids through the wounds. Some renal and severe liver patients may require less as previously indicated.
These values differ greatly depending on the reference. The early works by Kinney et al[4] and Long et al[7] are still in use today. A summary of some of the classifications of these two articles is as follows: Minor surgery 1.0 - 1.2 x BEE Major surgery 1.1 - 1.4 x BEE Skeletal trauma 1.2 - 1.35 x BEE Major sepsis 1.2 - 1.6 x BEE A modern summary of stress factors published by Frankenfield et al[19] is as follows: Trauma and major surgery in patients with a fever: 1.4 x BEE. Afebrile patients with similar injuries: 1.25 x BEE.	
For Burns:	**Protein[21]:**
There are many suggestions for estimating the energy required for burn patients. Some recommend 1.5 to 2.0 x BEE. A commonly used estimation of the energy requirement, and probably the more accurate one, is as follows[20]: Adults: [25 kcal x preburn body wt(kg)] + [40 kcal x %BSA burned] Children: [38 to 100 kcal* x preburn body wt(kg)] + [40 kcal x %BSA burned] * The actual kcals depends on the child's age.	2 to 3 g/kg of body weight per day A formula used to estimate protein loss per day is as follows[22]: Protein loss in grams per day = [UUN (g) + 4 + (0.2 g x %3° burn) + (0.1 g x%2° burn)] x 6.25 This formula is to allow for protein lost directly through the skin.

Table D - 6
Calculation of LDL Cholesterol

$$LDL = Cholesterol - HDL - \frac{Triglycerides}{5}$$

REFERENCES

1.   Harris, J.A. & Benedict, F.G. (1919) A biometric study of basal metabolism in man. *Carnegie Institute, Publication* 279:40.

2.   Food and Nutrition Board, National Research Council. (1989) *Recommended Dietary Allowances,* 10th Ed. Washington: National Academy Press.

3.   Weinsier, R.L., Heimburger, D.C. & Butterworth, C.E. (1989) *Handbook of Clinical Nutrition,* 2nd Ed. St. Louis, MO. The C.V. Mosby Company. pg. 179.

4.   Kinney, J.M., Duke, J.H., Long, C.L., & Gump, F.E. (1970). Tissue fuel and weight loss after injury. *J. Clin. Pathol.* 23(Supp 4):65-72.

5.   Whitney, E.N., Cataldo, C.B., & Rolfes, S.R. (1998) *Understanding Normal & Clinical Nutrition,* 5th Ed. Belmont, CA. West/Wadsworth. pg. 812-814.

6.   Lemon, P.W.R. (1995) Do athletes need more dietary protein and amino acids? *Int. J. Sport Nutr.* 5:S39-S61.

7.   Long, C.L., Schaffel, N., Geiger, J.W., Schiller, W.R., & Blakemore, W.S. (1979). Metabolic response to injury and illness: estimation of energy and protein needs from indirect calorimetry and nitrogen balance. *J. Parenter. Enteral Nutr.* 3(6):452-456.

8.   Weissman, C., Kemper, M., Damask, M.C., Askanazi, J., Hyman A.I., & Kinney, J.M. (1984). Effect of routine intensive care interactions on metabolic rate. *Chest.* 86(6):815-818.

9.   Swinamer, D.L., Phang, P.T., Jones, R.L., Grace, M., & King, E.G. (1987). Twenty-four hour energy expenditure in critically ill patients. *Crit. Care Med.* 15(7):637-643.

10. Frankenfield, D.C., Wiles, C.E. III, Bagley S., Siegel, J.H. (1994). Relationship between resting and total energy expenditure in injured and septic patients. *Crit. Care Med.* 22(11):1796-1804.

11. Whitney, E.N., Cataldo, C.B., & Rolfes, S.R. (1998). *Understanding Normal & Clinical Nutrition,* 5th Ed. Belmont, CA. West/Wadsworth. pg. 831.

12. O'Keefe, S.J., Abraham, R., El-Zayadi, A., Marshall, W., Davis, M., & Williams, R. (1981). Increased plasma tyrosine concentration in patients with cirrhosis and fulminant hepatic failure associated with increased plasma tyroseine flux and reduced hepatic oxidative capacity. *Gastroenter.* 81:1017-1024.

13. Cerra, F.B., Cheung, N.K., Fischer, J.E., Kaplowitz, N., Schiff, E.R., Dienstag, J.L., Bower, R.H., Mabry, C.D., Leevy, C.M., & Kiernan, T. (1985). Disease-specific amino acid infusion (FO[80]) in hepatic encephalopathy: a prospective, randomized, double- blind, controlled trial. *J. Parenter. Enteral Nutr.* 9:288.

14. Naylor, C.D., O'Rourke, K., Detsky, A.S., & Baker, J.P. (1989). Parenteral nutrition with branched chain amino acids in hepatic encephalopathy: a meta analysis. Gastroenter. 97:1033-1042.

15.   American Society for Parenteral and Enteral Nutrition Board of Directors. (1993). Guidelines for the use of parenteral and enteral nutrition in adult and pediatric patients. *J. Parenter. Enter Nutr.* 17(S 4):1S-53S.

16.   Renal Dietitians Dietetic Practice Group of the American Dietetic Association & the National Kidney Foundation Council on Renal Nutrition. (1993). National Renal Diet: Professional Guide. The American Dietetic Association.

17.   Zeman, F.J. (1991). *Clinical Nutrition and Dietetics*, 2nd Ed. Englewood Cliffs, N.J. Prentice Hall. pg. 314.

18.   Whitney, E.N., Cataldo, C.B., & Rolfes, S.R. (1998). *Understanding Normal & Clinical Nutrition*, 5th Ed. Belmont, CA. West/Wadsworth. pg. 944.

19.   Frankenfield, D.C., Smith, J.S., Cooney, R.N., Blosser, S.A., & Sarson, G.Y. (1997). Relative association of fever and injury with hypermetabolism in critically ill patients. *Injury.* 28(9-10):617-621.

20.   Curreri, P.W. (1979). Nutritional replacement modalities. *J. of Trauma* (supplement) 19:906-908.

21.   Bell, S.J. & Wyatt, J. (1985). Nutrition guidelines for burned patients. *J. of A. Dietetic Assoc.* 85:68-72.

22.   Weinsier, R.L., Heimburger, D.C. & Butterworth, C.E. (1989). *Handbook of Clinical Nutrition*, 2nd Ed. St. Louis, MO. The C.V. Mosby Company pg. 322.

23.   Tietz, N. Editor. (1976). Fundamentals of Clinical Chemistry. Philadelphia, PA. W.B. Saunders. Pg. 539.

APPENDIX E

M.V.I.®-12
MULTI-VITAMIN INFUSION[1]

DESCRIPTION

M.V.I.-12 consist of two vials labeled vial 1 and vial 2 in 5 ml single-dose vials and 50 ml multiple-dose vials.

Each 5 ml in vial 1 provides:
            ascorbic acid (vitamin C)..100 mg
            vitamin A* (retinol)..1 mg[a]
            ergocalciferol* (vitamin D)....................................5 mcg[b]
            thiamin (vitamin B_1)(as the hydrochloride)..............3 mg
            riboflavin (vitamin B_2)
            (as riboflavin-5 phosphate sodium)......................3.6 mg
            pyridoxine HCl (vitamin B_6)..................................4 mg
            niacinamide..40 mg
            dexpanthenol (d-pantothenyl alcohol)...................15 mg
            vitamin E* (dl-alpha tocopheryl acetate)................10 mg[c]

* Oil-soluble vitamins A, D, and E water-solubilized with polysorbate 80.
      a   1 mg vitamin A equals 3,300 USP units.
      b   5 mcg ergocalciferol equals 200 USP units.
      c   10 mg vitamin E equals 10 USP units.

Each 5 ml in vial 2 provides:

            biotin...60 mcg
            folic acid..400 mcg
            cyanocobalamin (vitamin B_{12}).........5 mcg

The contents of both vials should be added to not less than 500 cc of infusion fluid.

1 M.V.I.®-12 is manufactured by Centeon L.L.C., Kankakee, IL.
   All information presented here is from a M.V.I.®-12 product label.

M.T.E.®-4[2]
**MIXTURE OF FOUR TRACE ELEMENTS ADDITIVE
FOR IV USE AFTER DILUTION**

Ingredients	Single Dose Vial (3 ml fill)	Recommend Daily I.V. (TPN) Dose[3]
Zinc (as Zinc Sulfate heptahydrate)	1.0 mg	2.5 - 4 mg
Copper (as Cupric Sulfate pentahydrate)	0.4 mg	0.5 - 1.5 mg
Manganese (as Manganese Sulfate monohydrate)	0.1 mg	0.15 - 0.8 mg
Chromium (as Chromic Chloride hexahydrate)	4.0 mcg	10 - 15 mcg

2 M.T.E.®-4 is a product of Fujisawa USA, Inc., Deerfield, Il. All information presented here is from a product insert, revised September 1997.

3 Recommendation is for metabolically stable adults receiving TPN.

APPENDIX F

A PARTIAL LIST OF ENTERAL PRODUCTS

Product	Producer	Type	Form*	Cal/ml	Non-pro Cal/g N	Pro	g/L Cho	Fat	Na/L mEq	Na/L mg	K/L mEq	K/L mg	mOsm/kg Water	Vol to meet RDIs in ml	g of fiber /L	Free water /L in ml
Advera	Ross	AIDS	L	1.28	108:1	60	215.8	22.8	45.9	1056	72.5	2827	680	1184	8.9	802
AlitraQ	Ross	Elemental/ Glutamine	P	1.0	94:1	52.5	165	15.5	43.5	1000	30.7	1200	575	1500	N/A	846
Amin-Aid	McGaw	Renal	P	2.0	830:1	19.4	365.6	46.2	<5	<115	N¹	N¹	700	N/A	N/A	735
Choice	Mead Johnson	Diabetes/ Glucose Intolerance	L	1.06	123:1	45	106	51	37	850	47	1820	440	1000	14.4	850
Citrotein	Novartis	Cl Liq	P	0.67	76:1	41	120	1.6	29	670	14	550	480	1100	N/A	949
Compleat Modified	Novartis	Blenderized/ Fiber	L	1.07	131:1	43	140	37	43	1000	36	1400	300	1500	4.2	838
Compleat Pediatric	Novartis	Blenderized/ Fiber	L	1.0	142:1	38	126	39	29.6	680	38.5	1500	380	900	4.4	844
Comply	Mead Johnson	High Calorie	L	1.5	134:1	60	180	61	48	1100	11.8	463	460	830	N/A	775
Criticare HN	Mead Johnson	Elemental	L	1.06	149:1	38	220	5.3	27	630	34	1320	650	1890	N/A	830
Crucial	Nestle Clin Nutrition	Elemental	L	1.5	75:1	93.8	135	67.6	50.8	1168	48	1872	490	1000	N/A	768
Deliver 2.0	Mead Johnson	High Calorie	L	2.0	145:1	75	200	102	35	800	43	1700	640	1000	N/A	710
Diabeti-Source	Novartis	Diabetes/ Glucose Intolerance	L	1.0	100:1	50	90	49	40	930	36	1400	360	1500	4.4	849
Ensure	Ross	General	L	1.06	153:1	37.2	145	37.2	36.8	846	40	1564	470	1887	N/A	845
Ensure High Protein	Ross	High Nitrogen	L	0.95	92:1	50	128.3	25	52.5	1208	53.4	2083	610	960	N/A	846

* L = liquid, P = powder
N/A = not applicable or not available
¹ Negligible

Product	Producer	Type	Form*	Cal/ml	Non-pro Cal/g N	g/L Pro	g/L CHO	g/L Fat	Na/L mEq	Na/L mg	K/L mEq	K/L mg	mOs m/kg Water	Vol to meet RDIs in ml	g of fiber /L	Free water /L in ml
Ensure Light	Ross	Low Calorie	L	0.85	99:1	41.7	137.5	12.5	36.2	833	39.5	1542	475	960	N/A	816
Ensure Plus	Ross	High Calorie	L	1.5	146:1	54.9	200	53.3	45.7	1050	49.6	1940	690	1420	N/A	769
Ensure Plus HN	Ross	High Calorie/High Protein	L	1.5	125:1	62.6	199.9	50	51.3	1180	46.5	1820	650	947	N/A	769
Ensure with Fiber	Ross	Fiber	L	1.06	153:1	36.6	182.5	25.4	36.2	833	39.5	1542	500	948	15	812
Fiber-Source	Novartis	Fiber	L	1.2	149:1	43	170	39	48	1100	46	1700	490	1165	10	811
Fiber-Source HN	Novartis	Fiber/High Protein	L	1.2	115:1	53	160	39	48	1100	46	1700	490	1165	10	811
Glucerna	Ross	Diabetes/Glucose Intolerance	L	1.0	125:1	41.8	95.6	54.4	40.4	930	40.2	1570	355	1422	14.1	853
Glutasorb	Nutrition Medical	Elemental/Glutamine	L	1.0	96:1	52	186	6.7	26	600	25.6	1000	575	1800	N/A	732
Glytrol	Nestle Clinical Nutrition	Diabetes/Glucose Intolerance	L	1.0	114:1	45	100	47.5	32.2	740	35.9	1400	380	1400	15	850
Hepatic-Aid II	McGaw	Liver	P	1.2	148:1	44.1	168.5	36.1	<12.5	<288	<5	192	560	N/A	N/A	740
Immun-Aid	McGaw	AIDS	P	1.0	53:1	80	120	22	25	575	27	1055	460	2000	N/A	820
Impact	Novartis	Critically ill	L	1.0	71:1	56	130	28	48	1100	36	1400	375	1500	N/A	853
Impact 1.5	Novartis	Ctirically ill High Protein	L	1.5	71:1	80	140	69	56	1280	43	1680	550	1250	N/A	780
Impact with Fiber	Novartis	Critically ill Fiber	L	1.0	71:1	56	140	28	48	1100	36	1400	375	1500	10	868

* L = liquid, P = powder   N/A = not applicable or not available

Product	Producer	Type	Form*	Cal/ml	Non-pro Cal/g N	g/L Pro	g/L CHO	g/L Fat	Na/L mEq	Na/L mg	K/L mEq	K/L mg	mOsm/kg Water	Vol to meet RDIs in ml	g of fiber /L	Free water /L in ml
Introlite	Ross	Tube Feed Starter	L	0.53	125:1	22.1	70.6	18.4	40.4	930	40.2	1570	200	1321	N/A	920
Isocal	Mead Johnson	Isotonic Low Na	L	1.06	167:1	34	135	44	23	530	34	1320	270	1890	N/A	840
Isocal HN	Mead Johnson	Isotonic High Protein	L	1.06	125:1	44	123	45	40	930	41	1610	270	1180	N/A	850
Iso-Source	Novartis	Isotonic	L	1.2	149:1	43	170	39	48	1100	43	1700	490	1400	N/A	820
Iso-Source HN	Novartis	Isotonic High Protein	L	1.2	115:1	53	160	39	48	1100	43	1700	490	1165	N/A	820
Iso-Source VHN	Novartis	Isotonic High Protein	L	1.0	77:1	62	130	29	57	1300	41	1600	300	1250	N/A	847
Iso-Source 1.5	Novartis	Isotonic High Calorie	L	1.5	116:1	68	170	65	57	1300	54	2100	650	933	N/A	778
Isotein HN	Novartis	Isotonic/High Protein	P	1.19	86:1	68	160	34	27	630	28	1100	300	1770	N/A	859
Jevity	Ross	Isotonic/Fiber	L	1.06	125:1	44.3	154.7	34.7	40.4	930	40.2	1570	300	1321	14.4	835
Jevity Plus	Ross	High Protein/Fiber	L	1.2	110:1	55.5	172.7	39.3	58.7	1350	47.4	1850	450	1000	12	810
Kindercal	Mead Johnson	Children 1 - 10 Years	L	1.06	172:1	34	135	44	16	370	33.6	1310	310	V¹	6.3	850
L-Emental	Nutrition Medical	Elemental	P	1.0	149:1	38	205	2.8	20	460	20	780	630	2000	N/A	853
L-Emental Hepatic	Nutrition Medical	Elemental/Liver	P	1.2	148:1	44.1	168.5	36.2	<15	<345	<5	<195	560	N/A	N/A	820

* L = liquid, P = powder

N/A = not applicable or not available

V¹ = varies for each nutrient depending on the child's age.

Product	Producer	Type	Form*	Cal/ml	Non-pro Cal/g N	g/L Pro	g/L CHO	g/L Fat	Na/L mEq	Na/L mg	K/L mEq	K/L mg	mOs m/kg Water	Vol to meet RDIs in ml	g of fiber /L	Free water /L in ml
L-Emental Pediatric	Nutrition Medical	Elemental for Children 1 - 10 Years	P	0.8	200:1	24	130	24	17	400	31	1200	360	V¹	N/A	893
L-Emental Plus	Nutrition Medical	Elemental High Protein	P	1.0	115:1	45	190	6.7	27	610	28	1100	650	1800	N/A	833
Lipisorb	Mead Johnson	Fat Mal-absorption	L	1.35	125:1	57	161	57	59	1350	43	1690	630	1600	N/A	800
Magnacal Renal	Mead Johnson	High Calorie Hemo-dialysis	L	2.0	142:1	75	200	101	34.8	800	32.6	1270	570	1000	N/A	710
Meritene	Novartis	General	P	1.06	71:1	69	120	34	48	1100	72	2800	690	N/A	N/A	819
Nepro	Ross	Renal	L	2.0	154:1	70	222.3	95.6	36.7	845	27.2	1060	665	960	N/A	699
Nova Source	Novartis	High Cal/ High Protein	L	2.0	116:1	90	220	88	35	800	44	1700	790	948	N/A	701
Nova Source Pulmonary	Novartis	Respiratory Disease/ Fiber	L	1.5	102:1	75	150	68	56	1290	63	2300	650	933	8.0	764
Nova Source Renal	Novartis	Renal	L	2.0	140:1	74	200	100	43.5	1600	20.8	810	708	1000	N/A	708
NuBasics	Nestle Clinical Nutrition	General	L	1.0	154:1	35	132.4	36.8	38	876	32	1248	510	2000	N/A	850
NuBasics Plus	Nestle Clinical Nutrition	High Calorie	L	1.5	154:1	52	176	65	50.8	1168	48	1867	620-650	1333	N/A	780
NuBasics 2.0	Nestle Clinical Nutrition	High Calorie	L	2.0	131:1	80	196	106	56.6	1300	49.2	1920	750	750	N/A	700

* L = liquid, P = powder
N/A = not applicable or not available
V¹ = varies for each nutrient depending on the child's age.

Product	Producer	Type	Form	Cal/ml	Non-pro Cal/g N	g/L Pro	g/L CHO	g/L Fat	Na/L mEq	Na/L mg	K/L mEq	K/L mg	mOsm/kg Water	Vol to meet RDIs in ml	g of fiber/L	Free water/L in ml
NuBasics with Fiber	Nestle Clinical Nutrition	Fiber	L	1.0	154:1	35	132.4	36.8	38	876	32	1248	520	2000	14	840
NuBasics VHP	Nestle Clinical Nutrition	High Protein	L	1.0	75:1	62.4	112.8	33.2	38	876	32	1248	460	2000	N/A	850
Nutren 1.0	Nestle Clinical Nutrition	General	L	1.0	131:1	40	127	38	21.7	500	32.1	1252	300	1500	N/A	852
Nuten 1.0 with Fiber	Nestle Clinical Nutrition	Fiber	L	1.0	131:1	40	127	38	38.1	876	38.1	1248	310-370	1500	14	840
Nutren 1.5	Nestle Clinical Nutrition	High Cal	L	1.5	131:1	60	169.2	67.6	50.8	1170	48	1872	430-530	1000	N/A	780
Nutren 2.0	Nestle Clinical Nutrition	High Cal	L	2.0	134:1	80	196	106	56.6	1300	49.2	1900	720	750	N/A	700
Nutren Junior	Nestle Clinical Nutrition	Pediatric 1 - 10 Years	L	1.0	183:1	30	137.5	38.5	20	460	33.8	1320	350	1000	N/A	850
Nutren Junior with Fiber	Nestle Clinical Nutrition	Pediatric 1 - 10 Years	L	1.0	183:1	30	137.5	38.5	20	460	33.8	1320	350	1000	6	850
NutriHep	Nestle Clinical Nutrition	Liver	L	1.5	209:1	40	290	21.2	13.9	320	33.9	1320	690	1000	N/A	760
NutriVent	Nestle Clinical Nutrition	Pulmonary	L	1.5	115:1	68	100.8	94.8	33.9	780	32	1248	330-450	1000	N/A	780

* L = liquid, P = powder
N/A = not applicable or not available

Product	Producer	Type	Form*	Cal/ml	Non-pro Cal/g N	Pro	CHO	Fat	Na/L mEq	Na/L mg	K/L mEq	K/L mg	mOs m/kg Water	Vol to meet RDIs in ml	g of fiber /L	Free water /L in ml
						g/L										
Optimental	Ross	Elemental/Malabsorption	L	1.0	97:1	51.3	138.5	28.4	46	1060	45	1760	540-580	1422	N/A	835
Osmolite	Ross	Isotonic	L	1.06	153:1	37.1	151	34.7	27.8	640	26.1	1020	300	1887	N/A	841
Osmolite HN	Ross	Isotonic/High Protein	L	1.06	125:1	44.3	143.9	34.7	40.5	930	40.2	1570	300	1321	N/A	842
Osmolite HN Plus	Ross	High Protein/Cal	L	1.2	110:1	55.5	157.5	39.3	61.7	1420	49.7	1940	360	1000	N/A	820
Oxepa	Ross	ARDS	L	1.5	125:1	62.5	105.5	93.7	57	1310	50.3	1960	493	947	N/A	785
PediaSure	Ross	Pediatric 1-10 Years	L	1.0	184:1	30	109.7	49.7	16.5	380	33.6	1310	310	1000	N/A	844
PediaSure with Fiber	Ross	Pediatric 1-10 years Fiber	L	1.0	184:1	30	113.5	49.7	16.5	380	33.6	1310	345	1000	5	844
Peptamen	Nestle Clinical Nutrition	Elemental	L	1.0	131:1	40	127	39	24.3	560	38.5	1500	270-380	1500	N/A	850
Peptamen Junior	Nestle Clinical Nutrition	Elemental Pediatric 1-10 Years	L	1.0	183:1	30	137.5	38.5	20	460	33.8	1320	260-360	1000 N/A	N/A	850
Peptamen VHP	Nestle Clinical Nutrition	High Protein	L	1.0	75:1	62.5	104.5	39	24.3	560	38.5	1500	300-430	1500	N/A	840
Perative	Ross	Critical Care	L	1.3	97:1	66.6	177.2	37.4	45.2	1040	44.2	1730	385	1155	N/A	789
Pro-Balance	Nestle Clinical Nutrition	Fiber	L	1.2	114:1	45	130	33.8	27.7	636	33.3	1300	350-450	100	8.3	820
Promote	Ross	High Protein	L	1.0	75:1	62.5	130	26	43.5	1000	50.6	1980	340	1000	N/A	840
Promote with Fiber	Ross	Fiber	L	1.0	75:1	62.5	138.3	28.2	56.5	1300	50.8	1980	380	1000	14.4	830

* L = liquid, P = powder
N/A = not applicable or not available

Product	Producer	Type	Form	Cal/ml	Non-pro Cal/g N	g/L Pro	g/L CHO	g/L Fat	Na/L mEq	Na/L mg	K/L mEq	K/L mg	mOsm/kg Water	Vol to meet RDIs in ml	g of fiber /L	Free water /L in ml
PRO-Petide	Nutrition Medical	Isotonic/ Peptide	L	1.0	131:1	40	127	39	22	500	32	1250	270	1500	N/A	880
PRO-Peptide for Kids	Nutrition Medical	Pediatric 1 - 10 Years	L	1.0	183:1	30	137.5	38.5	20	460	34	1320	360	1000	N/A	850
PRO-Peptide VHN	Nutrition Medical	Isotonic High Protein	L	1.0	75:1	62.5	104.5	39	24	560	38.5	1500	300	1500	N/A	840
Protain XL	Mead Johnson	Wound Healing/ Fiber	L	1.0	86:1	57	129	30	40	920	45.1	1760	340	1250	9.1	830
Pulmocare	Ross	Pulmonary	L	1.5	125:1	62.6	105.7	93.3	57	1310	44.2	1730	475	947	N/A	785
Reabilan	Nestle Clinical Nutrition	Peptide	L	1.0	175:1	31.5	131.5	40.5	30.5	700	32.1	1250	350	2000	N/A	850
Reabilan HN	Nestle Clinical Nutrition	Peptide/ High Nitrogen	L	1.33	117:1	58	158	54	43	1000	42	1661	490	1500	N/A	810
Renalcal	Nestle Clinical Nutrition	Renal	L	2.0	338:1	34.2	290.4	82.4	N/A	N/A	N/A	N/A	600	1000[1]	N/A	700
Replete	Nestle Clinical Nutrition	HN	L	1.0	75:1	62.5	113	34	21.7	500	40	1560	290	1000	N/A	844
Replete with Fiber	Nestle Clinical Nutrition	Fiber	L	1.0	75:1	62.5	113	34	21.7	500	40	1560	300	1000	14	836
ReSource Diabetic	Novartis	Diabetes/ Glucose Intolerance	L	1.06	79:1	63	99	47	42	970	29	1100	450	1890	12.7	874

* L = liquid, P = powder
N/A = not applicable or not available
[1] = for water soluble vitamins only

Product	Producer	Type	Form	Cal/ml	Non-pro Cal/g N	g/L Pro	g/L CHO	g/L Fat	Na/L mEq	Na/L mg	K/L mEq	K/L mg	mOsm/kg Water	Vol to meet RDIs in ml	g of fiber /L	Free water /L in ml
ReSource Just For Kids	Novartis	Pediatric 1 - 10 Years	L	1.0	185:1	30	110	50	17	380	33	1300	390	1000	N/A	853
ReSource Select	Novartis	Glutamine/Fiber	P	1.3	73:1	42	85	27	29	667	12	467	N/A	N/A	10	873
ReSource Standard	Novartis	General	L	1.06	174:1	38	170	25	40	930	38	1480	650	946	N/A	840
ReSource Plus	Novartis	High Caorie	L	1.52	173:1	55	220	46	57	1310	50	1940	870	946	N/A	766
Resplor	Mead Johnson	Pulmonary	L	1.52	102:1	76	148	71	55.2	1270	37.9	1480	580	1420	N/A	770
SandoSource Peptide	Novartis	Peptide	L	1.0	100:1	50	160	17	52	1200	41	1600	490	1750	N/A	840
Suplena	Ross	Renal	L	2.0	393:1	30	255.2	95.6	34	790	28.7	1120	600	947	N/A	712
Sustacal Basic	Mead Johnson	General	L	1.06	152:1	38	146	38	37	850	40	1560	470	1890	4	840
Sustacal Liquid	Mead Johnson	High Protein	L	1.01	78:1	61	139	23	40	930	54	2100	650	1060	<4	850
Sustacal Plus	Mead Johnson	High Calorie/Protein	L	1.52	134:1	61	190	57	37	850	38	1480	630	1180	<4	780
Sustacal with Fiber	Mead Johnson	Fiber	L	1.06	120:1	46	139	35	31	720	36	1390	480	1420	10.6	850
Tolerex	Novartis	Elemental	P	1.0	282:1	21	230	1.5	20	470	31	1200	550	3160	N/A	864
Tramacal	Mead Johnson	Critical Care	L	1.5	91:1	82	142	68	51	1180	36	1390	490	2000	N/A	780

* L= liquid, P = powder
N/A = not applicable or not available

Product	Producer	Type	Form*	Cal/ml	Non-pro Cal/g N	g/L Pro	g/L CHO	g/L Fat	Na/L mEq	Na/L mg	K/L mEq	K/L mg	mOsm/kg Water	Vol to meet RDIs in ml	g of fiber /L	Free water /L in ml
Two-Cal HN	Ross	High Calorie High Protein	L	2.0	125:1	83.5	216.1	89.1	63.5	1460	62.8	2450	690	947	N/A	713
Ultracal	Mead Johnson	Fiber/High Protein/Isotonic	L	1.06	128:1	44	123	45	40	930	41	1610	310	1180	14.4	850
Vital HN	Ross	Elemental	P	1.0	125:1	41.7	185	10.8	24.6	566	35.8	1400	500	1500	N/A	867
Vivonex Pediatric	Novartis	Pediatric 1 - 10 Years/ Elemental	P	0.8	200:1	24	130	24	17	400	31	1200	380	1000	N/A	893
Vivonex Plus	Novartis	Elemental	P	1.0	115:1	45	190	6.7	27	611	27	1056	650	1800	N/A	850
Vivonex T.E.N.	Novartis	Elemental/ Glutamine	P	1.0	149:1	38	210	2.8	26	600	24	950	630	2000	N/A	853

* L = liquid, P = powder
N/A = not applicable or not available

APPENDIX G

Electronic Resources Related to Nutrition

Source	Electronic Address
American Cancer Society	http://www.cancer.org
American College of Clinical Nutrition	http://www.faseb.org/ascn/
American College of Obstetricians and Gynecologists	http://www.acog.com/
American Dietetic Association	http://www.eatright.org/
American Heart Association Food and Nutrition	http://207.211.141.25/
American Institute of Nutrition	http:www.nutrition.org/nutriton/
American Journal of Clinical Nutrition	http://www.faseb.org/ajcn/
American Medical Association	http://www.ama-assn.org
American Public Health Association	http://www.apha.org
Archives of Gynecology and Obstetrics	http://link.springer.de/link/service/journals/00404/index.htm
Caffeine	http://h-devil-www.mc.duke.edu/h-devil/nutrit/caffeine.htm
Canadian Eating Disorders Sites	http://www.stud.unit.no/studorg/ikstrh/ed/ed_cana.htm
Center for Disease Control & Prevention	http://wonder.cdc.gov/
Center for Food Safety and Applied Nutrition: Protecting Yourself Against Health Fraud	http://vm.cfsan.fda.gov/~dms/wh-fraud.html
Center for Science in the Public Interest	http://www.cspinet.org/
Choosing a Safe and Successful Weight-Loss Program	http://www.niddk.nih.gov/Saf&Suc/Saf&Suc.html

Clinical Nutrition On-Line	http://indigo.ie/~comdiet /Dietitian-Online.htm
Consumer's Guide to Fats	http://ificinfo.healthy. org/insight/carbo.htm
Cornell University Food Science	http://www.nysales.Corne ll.edu/cifs/
Creatine: A Review of Its Uses in Sports	http://wwwnetstorage.com/ hon/summary.html
Crohn's and ulcerative colitis site	http://www-medlib.med.ut ah.edu/WebPath/IBD.html
Crohn's Disease/Ulcerative Colitis Home Page	http://qurlyjoe.bu.edu/ cduchome.html
Dietary Guidelines for Americans	http://www.nalusda.gov./ fnic/dga/dguide95.html
Dietetics Online	http://www.dietetics.com
Duke University: Diet & Fitness Center	http://dmi-www.mc.duke. edu/dfc/home.html
Duke University: Alcohol and Health Page	http://h-devil-www.mc. duke.edu/h-devil/drugs/alcohol.htm
Fat-Free-USDA Nutrient Values	http://www.fatfree.com:80/usda/
Fat-Free: The Low Fat Vegetarian Archive	http:wwww.fatfree.com:80/index.sthml
Fat: How Much is Too Much?	http://h-devil-www.mc. duke.edu/h-devil/nutrit/fat.htm
FDA Consumer Magazine	http://www.fda.gov/fdac/796_toc.html
Federal Register	http://thorplus.lib.purdue.edu:80/gpo/
Federal Trade Commision	www.ftc.gov
Food and Drug Administration's Guide to Vegetarianism	http://fda.gov/opacom/catalog/vegdiet.html

Food and Nutriton Information Center (FNIC) Part of the National Agricultural Library

http://www.nal.usda.gov/fnic
or
gopher.nal.usda.gov

Food and Drug Admin. Home Page

http://www.fda.gov/fdahomepage.htm

Food Institute of Canada

http://foodner.fic.ca

Food Resource

http://www.orst.edu/food -resourse/food.html

Health Inspector

http://www.homeless.com. homepages/ael858@freenet. toronto.on.ca.html

Healthy School Meals Initiative/USDA

http://schoolmeals.nalusda.gov:8001

Hershey Foods Corporation

http://www.hersheys.com/hershey/

Information about Dietary Supplements

http://vm.cfsan.fda.gov/ ~dms/supplmnt.html

Institute of Food Technoligists (IFT)

http://www.ift.org/

Jean Mayer United States Department of Agriculture, Human Nutrition Research Center on Aging

http://www.hnrc.tufts.edu

Library of Congress

http://lcweb.loc.gov

Mayo Clinic

http://healthent.ivi.com/

Minority Health Resource Pocket Guide

http://www.omhrc.gov/pocket/pocket.htm

National Clearninghouse for Alcohol and Drug Information

http://www.health.org

National Council Against Health Fraud Home Page

http://www.primenet.com/~ncahf/

National Food Safety Database

http://www.agen.ufl.edu/~ foodsaf/foodsaf.html

National Institutes of Aging

http://www.nih/gov/nia

National Institutes of Diabetes
and Digestive and Kidney Diseases
(NIDDK)

http://www.niddk.nih.gov/
NIDDK_HomePage.html

National Institutes of Health

http://www.nih.gov

National Library of Medicine

http://www.nlm.nih.gov

National Osteoporosis Foundation

http://www.nof.org/

New England Journal of Medicine

http://www.nejm.org/

Nutrition and the Athlete from the
University of Nebraska Extension

http://www.ianr.unl.edu/
pubs/NebFacts/nf92-66.htm

Nutrition and Maternal/Child Health

http://weber.u.washington.edu/~larsson/
/phnutrit/internet/nutrlist.html

Nutrition Supplements

http://h-devil-www.mc.
duke.edu/h-devil/nutrit/suppl.htm

Nutrition and Your Health: Dietary
Guidelines for Americans

http://vm.cfsan.fda.gov/~
dms/nutguide.html

Online - Diet and Cancer

http://oncolink.upenn.edu/causeprevent/
diet/index.html

Patient Information Documents
on Nutrition and Obesity

http://www.niddk.nih.gov/
NutritionDocs.html

Penn State Sports Medicine Newsletter
Homepage

http://cac.psu.edu/~hgk2/index.html

Purdue University's extension
publications

http://pasture.ecn.purdue.
edu/~brendale/documents.html

Quackwatch: Your Guide to Health
Fraud, Quackery, and Intelligent
Decision-making

http://www.quackwatch.com

Step by Step: Eating to Lower
Your High Blood Cholesterol

http://www.nih.gov/news/

The American Anorexia/Bulimia
Association

http://members.aol.com/
amanbu/index.html

The American Society of Parental
and Enteral Nutrition (ASPEN)

http://www.peakcom.com/
clinnutr.org/